*Am I Dreaming?*

JAMES KINGSLAND is a science journalist with more than twenty-five years' experience working for publications including *New Scientist* and *Nature*. Most recently he was a commissioning editor and science production editor for the *Guardian*. He is the author of *Siddhartha's Brain: Unlocking the Ancient Science of Enlightenment*.

# *Am I Dreaming?*

## The NEW SCIENCE of CONSCIOUSNESS and HOW ALTERED STATES REBOOT the BRAIN

JAMES KINGSLAND

Atlantic Books
London

Published in hardback and trade paperback in Great Britain in 2019 by
Atlantic Books, an imprint of Atlantic Books Ltd.



1 3 5 7 9 10 8 6 4 2

A CIP catalogue record for this book is available from the British Library.

Hardback ISBN: 978 1 78649 550 1
Trade paperback ISBN: 978 1 78649 551 8
E-book ISBN: 978 1 78649 552 5

Printed in Great Britain Bell and Bain Ltd, Glasgow

Atlantic Books
An imprint of Atlantic Books Ltd
Ormond House
26–27 Boswell Street
London
WC1N 3JZ

www.atlantic-books.co.uk

*For our guardian angels,*
*Stefana, Biz, Amit and Brooks*

# *Contents*

# *Introduction*

*Is all that we see or seem, but a dream within a dream?*
Edgar Allan Poe, 'A Dream Within a Dream'

The shaman's assistant shone her torch in my face and whispered in my ear, 'Are you OK, James?' like a kindly nurse to a patient coming round after an operation. A French woman in her early thirties who spoke several languages fluently, her role in the ceremony was to translate the shaman's Spanish instructions, issue calm reassurance as required and usher us to the toilets in the dark. I told her I felt just fine. 'That's good,' she said. 'Do you want to drink again?'

The ink-black interior of the ceremonial hut, or *maloca*, raised on stilts over a muddy stream in the Peruvian Amazon did feel like a hospital ward. My fellow patients were sitting or lying on wipe-down plastic mattresses in the darkness on either side of me, each with a puke bucket within easy reach. The more organized among us had brought our own torches to light our way on the almost inevitable, urgent dashes to relieve ourselves that we were going to have to make during the night.

A whole range of afflictions had brought us to this jungle ward in February 2017. Some were seeking healing for drug addiction, depression or past traumas. Others, like me, simply yearned for a greater sense of meaning and purpose in their lives, a spiritual epiphany that mainstream religion had somehow failed to deliver, an antidote for the frustration and cynicism of middle age. For weeks, in preparation for the 'operation', we had abstained from sexual activity and followed a highly restrictive diet free from red meat, spices, salt and pepper, oils, animal fat, dairy, chocolate, carbonated drinks, tea, coffee and alcohol. At dusk, the female shaman had ritually cleansed us with wild tobacco smoke, blowing it onto the palms of our hands, the tops of our heads and into our clothes. Finally we were called forward one at a time to gulp down a personalized dose of the bitter medicine, known as ayahuasca or yagé. Once swallowed, there was no turning back. We were strapped in for what could be a frightening ride.

'Are you *sure* about this?!' had been the response of my former colleague Ian Sample, the science editor at the *Guardian* newspaper in London, when I emailed him a month earlier to tell him what I was planning. 'It sounds fun/terrifying/bonkers.' I was sure – at least at first.

Before the arrival of European settlers and Christianity in the late fifteenth century, ayahuasca was widely employed by the indigenous peoples of the Amazon in their religious ceremonies, in rites of passage and as a medicine. During the colonial era its use was suppressed and survived only in the relatively inaccessible Upper Amazon, but in the past decade plane-loads of Western tourists have descended on the region to drink the psychedelic brew, and in parallel there has been an explosion of scientific interest.

I travelled to Peru partly as research for a *Guardian* article but also in the hope of improving my own well-being. A few months before my adventure, I read a study suggesting that ayahuasca can change a person's outlook, making them less judgemental and emotionally reactive, improving their ability to stay mindful in challenging circumstances.[1] A few studies hinted that the hallucinogenic tea might also have antidepressant properties. Others suggested it could be used in conjunction with psychotherapy to treat addictions and post-traumatic stress disorder (PTSD).[2,3] To be honest, this all seemed a little too good to be true. I had interviewed some of the scientists behind the studies and now I wanted to try the medicine for myself.

After filling out a battery of medical questionnaires in which I painted a rosy picture of my physical and mental health, I was excited to be offered a place on an ayahuasca retreat at the highly-regarded Temple of the Way of Light near Iquitos in Peru. The temple's intensive treatment programme would involve five ayahuasca ceremonies in the space of nine days, personal consultations with facilitators, workshops and cleansing 'flower baths'. Reassuringly, the ceremonies would be conducted by experienced Shipibo healers with English-speaking facilitators on hand throughout.

A little over a week before my flight, however, I started to get cold feet. It wasn't the rare but widely reported fatalities among the thousands of Western tourists who had drunk ayahuasca in South America over the past few years that spooked me. (On closer inspection most, if not all, of these deaths turned out to be associated with poorly run retreat centres and caused by factors unrelated to ayahuasca itself, such as reactions to other psychoactive drugs and a road traffic accident.)[4] It was my family history of bipolar disorder.

I don't personally have the condition, which causes alternating bouts of crushing depression and mania verging on psychosis, but drinking ayahuasca or smoking its psychedelic component dimethyltryptamine (DMT) has been known to 'unmask' these symptoms in people who are genetically predisposed to develop either bipolar disorder or schizophrenia.[5] The same is true of all the classic psychedelics. When the psychedelic properties of DMT's hell-raising cousin lysergic acid diethylamide (LSD) were first identified in the forties, scientists were initially more intrigued by its ability to provoke symptoms of psychosis, such as hallucinations and delusions, than they were by its promise as a medicine. In a series of clandestine research projects in the US in the fifties and sixties as part of its MKUltra programme, the CIA investigated the possibility of using LSD as a mind-control weapon to temporarily scramble the brains of high-ranking enemy officials before important meetings or speeches, as a form of mental torture to elicit confessions from foreign agents, and to brainwash subjects into becoming 'robot agents'. The investigations were eventually abandoned after it became clear the effects of the drug were too unpredictable, but not before hundreds of unwitting subjects had been dosed without their consent.

LSD, DMT and psilocybin (the psychedelic component of magic mushrooms and truffles) are grouped under the title 'classic psychedelics' because they all achieve their transient, psychosis-like effects by binding to the same molecule in the brain, the serotonin 2A receptor. The receptor's normal binding fellow – serotonin – is a neurotransmitter that boosts nerve signal transmission, and there has been speculation that the receptor is involved in responses to extreme stress. Worryingly, some antipsychotic drugs appear to work by *preventing* serotonin from binding to the serotonin 2A receptor. In the days and weeks after the immediate effects of a classic psychedelic have worn off, the risk of psychosis or mania is very small, even among those like me who may be genetically vulnerable to these conditions. Nevertheless it has been estimated that around a third of the people who are unlucky enough to be affected in this way will go on to develop schizophrenia or bipolar disorder.[6] This was starting to scare me. Was I really prepared to gamble with one of my most precious possessions – sound mental health – for the sake of curiosity?

Foolishly, I deliberately hadn't mentioned my bipolar family history in the screening questionnaires for the Temple of the Way of Light. I had also failed to reveal an odd experience at university decades earlier when I felt wired for several days and nights for no apparent reason. Was that a manic episode, I wondered, a glimpse of a genetic chink in my mental armour? A dose of Valium prescribed by my GP at the time brought me safely back down to earth, so I hadn't thought much more about it. The experience never recurred, but the recollection was making me feel increasingly jittery about drinking ayahuasca.

When I did finally come clean about my family's history of bipolar disorder, the Temple of the Way of Light promptly withdrew my place on its programme. Mental breakdowns after ceremonies are rare, wrote the bookings officer in a friendly but firm email, 'but we have seen how difficult it can be to recover from psychosis for some of these folks, and we are very aware that we are not equipped with the professional psychological staff to safely support these individuals'. The nearest general medical clinic, she wrote, let alone a hospital, could only be reached by a two-hour hike through the jungle and a boat trip down the Amazon. My request to attend the ceremonies as an observer was also turned down, on the grounds that my presence might disrupt the shamans' delicate healing work with participants.

But my flight to Iquitos was already booked and I was determined to at least witness a ceremony at one of the other centres in and around the city. If the experience was sufficiently reassuring, I decided, I would screw up my courage and drink ayahuasca myself, on the condition that I received a relatively low dose and didn't have to commit to a mentally gruelling series of ceremonies. After three or four failed enquiries and just days before my flight, I found a reputable retreat centre about twenty miles from Iquitos that was prepared to accept me on these terms.

The Dios Ayahuasca Sanaciones healing centre turned out to be little more than a cluster of thatched huts in a clearing about half an hour's hike through the jungle from the highway. There was no electricity or running water, but the place was clean and well-maintained. The staff, though they spoke little English, were helpful and my fellow guests were friendly, relaxed and welcoming. My confidence was growing. Through the translator, I spoke to the shaman who runs the centre about my family's history of bipolar and asked if he thought it was a good idea to drink a small dose of the medicine at that evening's ceremony. Gazing at me intently for a few seconds as if he could read the stability of my mind in my eyes, he nodded and said yes, everything was going to be fine. So it was that eight days after being turned away from the Temple of the Way of Light, I found myself sitting on a mattress with my back to a wooden pillar in the centre's maloca, waiting for my first psychedelic trip to begin. I knew a little about the biochemistry of what was now happening inside my body. The genius of ayahuasca is that, in addition to DMT from the shrub *Psychotria viridis*, the brew contains chemicals from the vine *Banisteriopsis caapi* known as 'monoamine oxidase inhibitors', which disarm an enzyme that would otherwise break down the psychedelic before it could have any effect on the nervous system.

About half an hour after ingesting the foul-tasting liquid, I convinced myself I could feel the drug's hot, unstoppable progress through my body, from my guts into my veins and onwards to my brain, then spreading like a fire beneath my scalp. A burning drop of sweat ran down my brow and into one eye. As DMT took control of my senses, the nocturnal chorus of hoots, barks and growls in the surrounding jungle seemed to grow louder, answering the shaman's melancholic, enchanting *icaros* or medicine songs, which are said to summon the plant spirits. Directly behind me, close to the pillar where I sat, I distinctly heard the rhythmic clatter of a *chakapa*, a rattle made from a bundle of dried palm leaves. But when I turned my head there was nobody there. Regardless of the evidence of my eyes, however, the rattling continued.

My neighbour a few feet to my left, a man in his early twenties from Macedonia and a veteran of half a dozen ceremonies, reached for his bucket, dry-retched into it and giggled happily. I was beginning to feel a little nauseous myself, though I failed to see the funny side. Apart from vomiting and diarrhoea – which afflict nearly everyone – ayahuasca rookies are often gripped early on in their trip by overwhelming terror. Having your ego chemically stripped away can feel like the annihilation of death. 'Surrender yourself to the experience,' I'd been advised a few weeks earlier by an experienced user. If you don't fight the drug, sensations of extraordinary bliss and peace may follow; vivid visions of exotic rainforest creatures, healing encounters with the plant spirit Mother Ayahuasca, mind-blowing adventures.

Waiting for the ceremony to start about an hour earlier, my other neighbour – a Londoner in his thirties – had reminisced about a trip the previous year during which he roared into the night sky over the jungle in the cockpit of a space shuttle. Looking down at the ceremony in the rapidly receding maloca far below, he saw pyramids erupt through gaps between the floorboards. My own visions, after a more modest dose of the medicine, were rather less dramatic. When I closed my eyes I found myself on a balcony in a colonial-style monastery overlooking a cloistered courtyard awash with seething, brightly coloured geometric shapes. But what filled me with joy and wonder – the thing that really sticks in my memory – was the shaman's plaintive song; its volume, beauty and ineffable *meaningfulness* magnified by ayahuasca.

Some time later – I thought it must be nearly dawn but it turned out only a few hours had passed – and the effects were starting to wear off. There had been no terror, no ego dissolution, only an awestruck fascination with the whole perception-warping, magical experience. I was nauseous and my gut was rumbling but I hadn't felt the need to use my bucket or dash to the toilet since downing the brew. So when the shaman's assistant approached to ask me if I wanted to drink again, I was tempted. Then I remembered the small but real risks for people like me with a family history of bipolar disorder or schizophrenia and decided that enough was enough for now. I didn't want to push my luck and so I declined. Almost immediately I regretted not diving deeper into the extraordinary realms of consciousness I had heard others describe. I was left with a nagging sense that my life would have been richer for the experience. In the years following this first, tentative experience I have embarked on bolder adventures, with happy results, some of which are described later in this book.

To ingest a psychedelic drug is to take a leap of faith. Nobody can tell you in advance what will happen in that strange inner world after everyday reality has been suspended, or what the enduring consequences might be. Put like this it sounds frightening, and yet we make a similar leap into the unknown every time we close our eyes to sleep. Who knows what nightmares may come? There are other similarities. Like tripping with your eyes tightly closed, dreams are almost completely isolated from external, sensory reality and are mostly visual. The 'hypnagogic' visions of random, abstract shapes that people sometimes report seeing as they fall asleep recall the geometric patterns often witnessed during a trip. In both dreaming and tripping, time perception is distorted and, like those that happen during a trip, the bizarre narratives and encounters of our dreams are almost always first-person, subjective experiences – quite unlike watching a film or TV drama. Could the same underlying mechanism explain the biological purpose of dreams and the therapeutic promise of psychedelics?

Sigmund Freud, the father of psychoanalysis, argued that dreaming provides a safe outlet for fulfilling repressed sexual desires, and that by interpreting dreams a skilled therapist could bring these desires to light and effect a cure. Rather than uncovering repressed urges, neuroscientists now believe it is the occasionally unnerving suspension of sensory reality checks that occurs during all altered states – from dreams and hypnosis to psychedelics and deep meditation – that unlocks their potential benefits. In the process, however, they reveal a truth about ordinary consciousness every bit as unsettling.

Altered states of consciousness are temporary deviations from our normal, 'baseline' waking state involving multiple changes in perception, cognition, emotion and arousal levels. They may occur spontaneously, for example as a result of trauma, an epileptic fit or near-death experience, or they can be deliberately induced by drugs or practices such as sensory deprivation, fasting, breathing techniques or focused awareness. Regardless of their cause, by loosening the normal sensory and cognitive restraints, altered states can result in a breakdown of long-established beliefs about what is likely or unlikely, probable or improbable. They can even dissolve the deeply entrenched distinction between 'self' and everything else.

Dreams are the archetypal, everyday altered state that everyone has experienced, but what exactly are they for? Before you were born, dreams set the stage for your entrance into the world. If your mother was given an ultrasound scan at thirty weeks' gestation, it would have revealed your almost continuous rapid eye movements, or REMs, characteristic of dreaming sleep. What you dreamed in the warm darkness of the womb is anyone's guess, but your brain was almost certainly teaching itself two vital skills. First, as you kicked out and clenched your fists (unlike later in childhood and adulthood, your muscles still worked in your dreams), you were learning what it is to be an active agent situated in a physical body, known as 'core selfhood'; and second, as your eyes darted about behind closed eyelids – as if following the action in some hidden drama – you were taking your first lessons in how to see.

Once veiled in superstition, we now know that dreams play a crucial role in wiring highly adaptable brains, not only in humans but also in most other mammals and young birds. This is probably why human foetuses, infants and children dream so much. In adults, dreams are not only important for consolidating the memories needed to perform unfamiliar, complex tasks for the first time, but also for emotion regulation and creativity, as we shall see.

That dreams are a virtual-reality training ground for waking activities may not come as a surprise, but what if I were to tell you that even as you sit reading this, everything you see, hear, taste, smell, feel and touch is also only virtually real? The words on the page, the feel of the book or e-reader in your hands, perhaps the sound of distant traffic or conversations, the sense of your body occupying a particular space in a particular posture – none of these experiences arises directly from the data gathered by the photoreceptors in your eyes, the touch receptors in your fingertips, the microscopic hairs in your inner ears and the 'proprioceptors' recording the position and movement of your muscles. The weight of evidence now suggests they are virtual realities conjured by the brain using the same neural machinery that it uses to make your dreams.[7] According to this unnerving new perspective, rather than passively building a faithful, inner representation of the external world, the brain is constantly trying to stay one step ahead of the game, drawing on its past experiences to *predict* what's happening. Sensory information is not disregarded, but is relegated to the role of reality-testing the brain's guesswork and, as we shall see, it isn't necessarily given much credence.

The virtual nature of perception helps to explain the host of self-deceptions, sensory illusions and hallucinations to which we are prey. Why else would these distortions of reality, like dreams, seem so perfectly convincing? Having painstakingly scanned our own brains, recorded their electrical activity and scrutinized their constituent nerve cells, humans face a realization almost as disorienting as that faced by Keanu Reeves's character, Neo, in *The Matrix* as he watches a spoon held in a girl's fingers wilt before his eyes – then spring back upright. This is no cheap magic trick pulled off using subterfuge and phoney cutlery. To perform it, Neo is told, you must only realize the truth. 'There is no spoon,' the girl explains. 'It is not the spoon that bends, but only yourself.'[8]

Like Neo, it's time we came to terms with the discovery that the mind plays a leading role in everything we feel, see, hear, smell, taste and touch. This isn't to say that there is no objective, external reality, but our conscious representations of it are the product of the brain's innate virtual-reality generator. Neuroscientists now believe that in visual perception, for example, what we see is not the result of a three-dimensional, internal representation that our sensory cortex has laboriously built from the bottom up, step by step, by detecting features such as edges, lines and blobs in the raw sensory data, but effectively the *idea* of a spoon – the concept of a spoon that encompasses everything we know and have ever experienced in the world of spoons. We begin assembling our perceptual concepts from scratch in the womb and in infancy, collecting multisensory associations and committing them to memory. As a baby being weaned off milk and building a concept of spoons, among other things you learned to associate these objects' visual characteristics with those of food and your parents, with how the objects felt in your mouth, the taste of the food and the sensation of hunger satisfied. But the more you learned the less you relied upon raw, sensory spoon data and the more your consciousness drew upon the internal, virtual spoon.

As the brain develops, perceptions start to look less like direct sensations and more like predictions informed by context and similar past experiences, like templates held up before the mind's eye in order to judge how well they match information streaming from the senses. The job of the brain, it seems, is to minimize any discrepancies or 'prediction errors' by selecting the template that is the best match, perhaps updating it to further reduce future mismatches.

Needless to say, all this happens unconsciously and at lightning speed, but if we could run through the process in slow motion it might look something like this. Imagine you hear a knock on your front door. You open it and see – what? Your brain uses the context (time of day, whether you're expecting someone to drop by, and so on) to bring up the most probable templates: the postman, a friend, a neighbour, a complete stranger. The template it settles on will be the one that minimizes prediction errors, the differences between each available template and the limited sensory data on offer. It's the postman! Even so, a glaring visual-error signal remains. He has grown a beard, so your brain updates its 'postman' template accordingly.

This usually works perfectly well: not only does it save a lot of time and processing power, it also resolves the tricky problem that sensory information is inherently sparse, fuzzy and unreliable. We can never know what's happening in the world directly; we only infer it from context and the available sense data. Just as you draw upon your past experience to judge the veracity of the stories when you browse a news website (because journalists can't always be relied upon to report the news accurately and impartially), the brain must arbitrate between what it has learned previously – what it thinks it knows – and fresh sources of information. Instead of placing all its trust in meagre, noisy data, consciousness is founded upon prediction and expectation. The downside is that perceptions are easily bent out of shape – much more easily, in fact, than bending a spoon without touching it. As the girl in the film says, 'That would be impossible.'

Occasionally, when the brain slips up, its perceptual guesswork becomes glaringly apparent. Anyone who has ever stared out of the window of a moving train as the landscape or cityscape races past will be familiar with illusions of movement when the train comes to a halt. I remember my astonishment as a child when I first looked down and saw the ballast streaming alongside the tracks after our train had come to a halt at a station. Of course it was not the stones that were moving but my mind, which had become accustomed to expect this rapid flow. It took twenty seconds or so for the sensory data from my eyes to correct this impression, updating my brain's predictions about what was really happening.

Most of the time perceptual distortions like this hide in the shadows of ordinary consciousness, but they can be manipulated and enhanced at will. For thousands of years, humans have exploited the virtual nature of perception to transport themselves into alternative realms of experience. Among the powerful tools we have developed to twist our brains' predictions are hypnotic suggestion; trance states induced by dance, music or drumming; spiritual disciplines such as fasting, isolation, sleep and sensory deprivation, meditation and breath control; and, of course, psychedelic drugs. In thrall to these mind-bending influences, people are easily convinced that a spoon is wilting before their eyes, that black is white, that they are communing with their ancestors, with gods or angels, alien beings or the spirits of plants and animals. They may come to believe they can see into the future and resolve all manner of difficulties. And, perhaps most extraordinarily of all, they may shed their 'ego' or sense of autobiographical selfhood and perhaps even their bodily identity, melting distinctions between them and their surroundings, suddenly discovering a sense of union with everyone and everything else in the universe.

We have come to know these vivid, phantasmagoric experiences as altered states of consciousness, but they tell us much more about everyday consciousness than we might care to admit. They are powerful demonstrations of the endlessly creative, virtual nature of everything we have ever thought, felt or perceived.

Altered states are also proving invaluable tools for 're-tuning' the mind, adjusting the relative influence of fresh sensory experiences and established thought patterns and behaviours. They have enormous untapped potential for humanity. As we will see, psychedelics and meditation, in particular, are likely to be game changers in the coming years for people with intractable illnesses characterized by inflexible, destructive mental states, such as treatment-resistant depression, PTSD and addiction.

Around a decade ago, like so many others entering the middle years of their life, I began to ask myself: Is this it? On the surface I had little to complain about. I was happily married, living in a spacious house with a garden in a London suburb. I had a solid career in science journalism behind me and stretching ahead over the twenty-five or so years remaining before I could pay off my mortgage and retire. At forty-two years of age, I was earning a decent wage plugging away at a computer keyboard on the science desk of the internationally-renowned *Guardian* newspaper. I had climbed as high as I was likely to get on the journalistic career ladder, having started out in my twenties as a sub-editor on freebie magazines and newspapers for medical professionals, followed by a ten-year stint in my thirties as a sub-editor, editor and later writer for *New Scientist* magazine. With many of my youthful ambitions and passions satisfied, comfortably adapted to the ecological niche I had made my home, all that remained was to live out my life as a finished, fixed 'me'. This was not a prospect I relished, however. I did not like being me very much.

To give you a flavour of the person I was becoming, every weekday morning I would pedal to work on my bicycle, a journey of about five miles from Kilburn in north London to the *Guardian* offices in King's Cross, arriving forty minutes later in a mood fit to kill someone. During the ride everything and everyone would become steadily more hateful. It didn't help that I was usually running late, so every minor delay, from traffic lights turning red to buses stopping in front of me to pick up passengers, wound my mental springs a little tighter. The breaking point often came on Abbey Road in St John's Wood at the zebra crossing made famous by the Beatles, where every morning dozens of foreign tourists lined the pavement waiting their turn to be photographed, frozen in mid-stride like their heroes, oblivious to the blaring horns of the waiting rush-hour traffic. I remember one day I was so riled by this selfish behaviour that I didn't stop, shooting straight over the crossing ringing my bell furiously, barely missing a startled young man.

It was incidents like this that finally drew my attention to the person I was turning into. Rather than growing in wisdom with the passing years, I seemed to be getting angrier and more cynical, and while my political leanings were supposedly as liberal as they had been in my twenties, I realized that, deep down, I was becoming much less tolerant of others' behaviour and beliefs, who they were and where they came from. I also lacked any clear idea of what I wanted to do with the rest of my life. It was around this time that a Buddhist friend gave me a copy of *Mindfulness in Plain English* by the Sri Lankan monk Henepola Gunaratana. Maybe my friend had spotted those angry red flags? The book was my introduction to the Eastern idea that, through training, the mind can be persuaded to run along happier, calmer, more compassionate tracks, not simply loaded up with more and more information, which for a long time has been the Western way of thinking. At first I was sceptical, but I later discovered there was a growing body of scientific evidence suggesting that practices as simple as focusing on the breath or repeating a mantra have measurable physiological effects and lead to clear-cut changes in the brain, for example in areas involved in attention and emotion regulation.

I began to work in earnest on changing my mind, meditating first thing every morning and trying to be more mindfully aware from moment to moment of my surroundings and emotional state, not least on my journey to and from work. I was becoming calmer, but I also noticed more subtle effects on my sense of selfhood. I started to question my personal perspective on the world. On rainy days when I took the train rather than cycling, in the midst of the tide of humanity flowing out of King's Cross station my mind boggled at the sheer multiplicity of these other conscious 'selves', each implicitly assuming him or herself to be at the centre of the universe. They couldn't all be, could they?

Eventually I plucked up the courage to pack in my safe job at the *Guardian* in order to write *Siddhartha's Brain*, a book about Buddhism and the neuroscience of meditation. A few years later, my interest in psychedelic drugs was sparked when I read research suggesting that, after the acute, hallucinogenic effects of drinking ayahuasca have worn off, mindfulness is temporarily boosted to levels usually only seen in people who have been doggedly practising meditation for seven years or so.[9] I was intrigued. Given the chance, who in middle age wouldn't want to become a little less judgemental and emotionally reactive, to roll back some of the stultifying effects of the passing years? Mindfulness already figured large in my life. Could drinking an infusion brewed from the leaves of the shrub *P. viridis* and the mashed-up stem of the vine *B. caapi* really offer a shortcut to 'enlightenment' that didn't involve setting aside time every day to sit motionless on a cushion, year in and year out? I began to interview scientists who were researching psychedelics and other altered states, and my personal explorations eventually led to the ceremonial hut in the Peruvian jungle where I drank ayahuasca, and several months later to a darkened hotel room in Amsterdam where I ate psilocybin-containing 'magic truffles' for the first time. None of this prepared me, however, for the shock of ego death I experienced one hot afternoon in the Netherlands in May 2018, which I describe in Chapter 7.

Within these pages I explore a new scientific understanding of consciousness developed over the past decade – a grand unified theory that explains how human thought, emotion, perception and behaviour emerge from our brains' perpetual search for certainty in an uncertain world. In humans this search is also a quest for meaning. As I hope this book will demonstrate, wisely used, altered states are fabulous tools for widening the scope of our quest, restoring the broad, healthy perspective that mental illness, addiction and the passing years can take away. Like a zoom lens, they reveal the bigger picture that we so easily lose sight of the more closely our minds focus on the minutiae of everyday life. They can expand our consciousness to include everyone and everything.

By their very nature, however, altered states are risky. To adjust our focus, they temporarily suspend our sensory and cognitive reality-checking faculties, an issue I confront in Chapter 1. Paradoxically, one of the potential benefits of expeditions into altered states of consciousness may be the fine-tuning of these very faculties, helping us to become more mindful or lucid, even in our dreams (Chapter 2). In Chapter 3, I sing the praises of escapism through virtual-reality technology and video gaming which, contrary to tabloid wisdom, may offer considerable benefits for mental health. Virtual reality and games are also helping to reveal how the brain creates the boundaries of bodily selfhood. In Chapter 4, Balinese dancers and revellers demonstrate how extraordinarily easy it is to slip the anchors of selfhood through trance or hypnosis, allowing humans to assume the identities of gods, farm animals… and kitchenware.

A book about altered states of consciousness would be incomplete without a nod to the genius of Albert Hofmann, the creator of LSD, who from the start recognized the chemical's promise, not only in psychotherapy but also as a key for unlocking the secrets of consciousness. Fifty years after scientific work on LSD and the other classic psychedelics was effectively banned in the seventies, we are witnessing a renaissance in psychedelic research that has yielded the unexpected insight that the brain is exquisitely poised between order and chaos, stability and flexibility (Chapter 5). There may be even bigger surprises to come, not least if a controversial claim from an Australian PhD candidate – that a particular blend of ayahuasca tea, drunk under strictly controlled conditions, can lift the suicidal depression of bipolar disorder – is proved correct (Chapter 6).

One thing is certain. No matter how spectacular or awe-inspiring our experiences during altered states of consciousness, the insights gained must somehow be integrated into our everyday lives (Chapter 8). Maintaining a daily meditation practice may be particularly effective in this regard, though few people will become as skilled at observing their own mind as Tibetan monks, who can maintain conscious awareness even during non-REM, dreamless sleep, which they use as a training ground for their transition from this life to the next (Chapter 9).

In addition to exploring the rapidly developing neuroscience of consciousness, my hope is that this book will also provide inspiration for readers who want to dip their own toes in altered states with a view to widening the scope of their consciousness – not necessarily through drugs but perhaps through less daunting, gentler techniques such as lucid dreaming and self-hypnosis. Self-hypnosis is a proven technique for overcoming anxieties about situations such as dates, public speaking and job interviews; and lucid dreaming, as well as being great fun, can help see off our nightmares. You'll find step-by-step guides on how to attain each of these states on pages 57 and 110, respectively.

Over the years, like most humans, I have tinkered with my consciousness in countless ways. I have lost myself in books and music. As a teenager I loved playing the fantasy role-play game *Dungeons & Dragons* and, as an adult, I have dabbled in the virtual-reality *World of Warcraft*. I have been hypnotized in the hope that it would erase my lifelong fear of flying. I have meditated, taken all manner of prescription and non-prescription drugs, and trained myself to dream lucidly. None of these experiences fully prepared me for the lurch in consciousness wrought by a high dose of psychedelic that sunny afternoon in the Netherlands. More than any other altered state, these extraordinary drugs lay bare a mind in which our thoughts, beliefs, ideas and expectations play the starring roles in perception and selfhood.

# I

# *Magical Thinking*

It seems that as humans, even when all our immediate biological needs have been met, there remains a hunger for *meaning*. We long for a greater sense of connectedness with nature and the universe, some kind of reassurance that each of us plays a part in a grand narrative that will continue long after we have left the stage. Throughout recorded history, altered states of consciousness – whether they are brought on by drugs, music, drumming, dance, or exacting spiritual disciplines such as meditation, isolation and fasting – have satisfied this profound hunger, a yearning that has nothing to do with the basic biological drives for food, shelter, social status or sex. What other animal seeks out meaning?

This search for meaning is not an idle pursuit. To discover that you play a minor role in a cosmic drama that transcends your everyday self can be hugely beneficial for mental well-being. The impressive clinical effects of psychedelics recorded in studies over the past few years appear to be directly related to their ability to provoke meaningful, even spiritual experiences. The self-transcendent insights they afford have proved particularly helpful for people forced to confront their own mortality. A few years ago, when Stephen Ross and his colleagues at New York University School of Medicine gave a single dose of psilocybin (found in magic mushrooms and truffles) to twenty-nine people struggling to cope with a cancer diagnosis, the psychedelic significantly improved their quality of life and brought immediate and substantial relief from symptoms of anxiety and depression.

Their study employed a 'crossover' design: patients either took psilocybin in a first session followed by a placebo seven weeks later, or vice versa. Each was also given conventional psychotherapy. Six-and-a-half months after the treatment, between 60 and 80 per cent were still reporting clinically significant improvements in symptoms of anxiety and depression compared with the start of the trial. Crucially, this therapeutic effect seemed to be mediated by the spiritual insights they had while on psilocybin. Some 70 per cent of all the patients reported that, even though their trip was emotionally challenging, they rated it as among the top five most personally meaningful experiences of their entire lives.[1]

One of the patients, a fifty-one-year-old woman called Erin, who had been knocked sideways by the news she only had a 50:50 chance of being alive in five years' time as a result of ovarian cancer, described a vision that gave her a vital insight.[2] Under the influence of psilocybin, she saw a round dinner table:

> … and at the table was cancer, but it was *supposed* to be at the table. It isn't this bad, separate thing; it's something that's part of everything, and that everything is part of everything. And that's really beautiful. It was just a sort of acceptance of the human experience because it's all supposed to be this way.

She realized that cancer and death must have a place at the table: they are as integral to the natural order as life itself. And with that realization came the peace of acceptance.

Similarly, in a trial led by Robin Carhart-Harris at Imperial College London, reported in *The Lancet Psychiatry* in 2016, twelve patients with treatment-resistant major depression experienced significant, sustained reductions in their symptoms after taking just two doses of psilocybin one week apart alongside conventional psychotherapy. Again, these clinical improvements correlated with ratings of how insightful or mystical they judged the drug experience to be.[3–5]

In the modern world, doctors have assumed the healing role once played by priests and shamans, but theirs is a mostly biological conception of illness that can overlook patients' spiritual needs. Their pills are not designed to restore a sense of meaning, connectedness or purpose to people's lives. Prozac and Valium don't work by offering insights into one's problems but by damping down their emotional impact, which goes a long way towards explaining why conventional antidepressants and anti-anxiety drugs like these must be taken continually for their effects to be sustained, whereas research to date suggests that just one or two doses of a psychedelic can be life-changing. Also worth bearing in mind is the fact that taking conventional antidepressant and anti-anxiety drugs is often associated with side effects such as drowsiness and sexual dysfunction, and can lead to dependence: when patients stop taking them there may be unpleasant withdrawal effects, including a sharp rebound in their original symptoms.

Age-old techniques for provoking altered states of consciousness, such as meditation, sleep deprivation, trance and hallucinogens, are well known for their ability to precipitate spiritual and emotional breakthroughs but, as I would be the first to acknowledge, they are inherently risky for some people. There is a delicate balance to be struck between laying oneself wide open to spiritually meaningful experiences and triggering a 'psychotic episode' in which one can lose touch with reality for days, weeks or longer, with hallucinations and possibly delusions of persecution or grandeur. Research suggests that consciousness-warping chemicals and practices allow us to dissolve rigid patterns of thought and behaviour, including drug addiction and depression, like shrugging off old clothes that have grown worn and uncomfortably tight over the years. In the process, however, they expose the naked psyche to the cold blast of painful memories and emotions. And there can be no guarantee that the new clothes the mind finds to put on will fit any better than the old ones. For a few unlucky people, in particular those vulnerable to psychosis, the fit may be more uncomfortable.

All the altered states of consciousness I explore within these pages involve what psychologists call 'dissociation' – a temporary disconnection from everyday reality. But for the small percentage of people who have experienced psychosis or who have a family history of bipolar disorder or schizophrenia, these states run the risk of severing this link for longer periods, perhaps permanently. Psychedelic researchers recruiting volunteers for their studies will reject applicants if they fall into these categories, as will many meditation and ayahuasca retreat centres, as I learned to my disappointment. For everyone else, a supportive environment – a calm, protected space with compassionate, experienced individuals on hand – is essential to maximize the benefits and minimize the risks associated with these extraordinary experiences, in particular allowing any difficult emotions and memories that come up to be processed safely. Just as importantly, the healing process doesn't end when people return to ordinary consciousness. In the weeks and months that follow, the profound insights and revelations must somehow be integrated into everyday life, for example through a daily meditation practice, spending more time in nature, creative pursuits or voluntary work.

If you are wary of the idea of temporarily disengaging from reality, it's worth remembering that this is what happens every night while you sleep. In the next chapter, I explore the idea that dreaming sleep streamlines our models of the waking world, pruning redundant synapses that have accumulated during the day's learning experiences and helping our brains to function more efficiently. In order to do this, they must switch off almost all sensory inputs and motor outputs and suspend the mind's reality-checking faculties.

No wonder dreams, like the delusions of psychosis, are so very convincing. But if this nocturnal housework isn't carried out properly every night, the nervous system becomes increasingly cluttered with unnecessary connections. Like an ageing desktop computer, this means it works less efficiently, disrupting not just cognition but also homeostasis – the maintenance of a stable internal environment including essential tasks such as regulating temperature, pH and blood sugar levels. Not getting enough sleep (at least seven hours a night) is strongly associated with an increased risk of poor mental and physical health, including depression, suicidal thoughts, cancer, diabetes, heart disease and Alzheimer's.[6,7]

The effects of sleep deprivation on homeostasis are particularly striking. While research ethics committees would not allow such an experiment to be performed on humans, rats deprived of sleep for more than eleven days die as a result of a complete breakdown in their ability to maintain a stable body temperature.[8]

One of the take-home messages of this book is that you can't reap the restorative benefits of sleep, or any other altered state of consciousness, without temporarily disconnecting from the reality checks that your senses and rational mind normally provide. According to the leading theory of how we develop and maintain our cognitive models of the world, known as 'prediction error processing', the streamlining that underpins the restorative powers of altered states can only occur after entire levels of the brain's processing hierarchy have been taken offline.

The trouble starts when people mistake what they dreamed, or the visions they saw, for reality. Psychiatrists call this 'magical thinking'. Like many people, I have come to believe that science is our best friend for judging what is and is not real, so I get uncomfortable when some folk who have experienced altered states talk earnestly about supernatural phenomena such as astral projection – the ability to shed one's physical body and wander the cosmos – past lives and spirit guides. Like the contents of a dream in the moments after awakening, these things can seem all too real. Not even scientists are immune to magical thinking. Shortly before I travelled to Peru, I interviewed a highly respected researcher who has investigated the potential clinical benefits of drinking ayahuasca and personally taken part in many ceremonies. The changes in his worldview apparently wrought by the medicine were startling:

> Everything has consciousness. Plants, animals, rocks, you name it. The issue is how compatible is their awareness with our awareness. And in the case of plants they don't think. Their awareness is very, very different from ours. But when we reach out to the plant world and acknowledge that they feel, that they have awareness, plants will create a kind of hybrid awareness. These are the divas that folks in the shamanic realm will talk about.
>
> Ayahuasca has created a bridge to us, and this bridge is the spirit you will encounter in your ceremony. She is an absolute hard ally in this process. She will help you. If you develop a relationship with her and trust her she will take

> you exactly where she understands you need to go to heal, and that will be totally different from where your mind will think to go.

This well-meaning scientist's words only served to heighten my reservations about drinking ayahuasca. His conception of the medicine as opening up a channel of communication with plant spirit guides is in accord with what the shamans will tell you, and is a useful metaphor for how the healing works, but many regular users come to believe in a literal Mother Ayahuasca. Who knows? There may be something to their insight of a hidden realm of plant and animal spirits willing and able to help out with our problems. Personally I prefer to remain sceptical about such things.

The paradox of altered states of consciousness is that they restore openness, flexibility and meaning to our lives by temporarily messing with our reality-checking faculties. What fascinates me is that in the process, they reveal so much about how the brain strives to make sense of the world. The more we have learned about altered states, the more light they have shone on ordinary waking consciousness, exposing the hidden hand of belief, expectation and brain chemistry in everything we experience.

Optical illusions demonstrate that, as the British psychologist Richard Gregory observed in 1980, perceptions are like scientific hypotheses.[9] They are our best guesses based on the limited evidence available to us. Because sensory stimuli are inherently ambiguous, the brain must weigh the rough-and-ready 'rules of thumb' it has learned (its expectations based on past experience) against the likelihood that the available sense data is reliable. The result of this subconscious calculation is conscious perception. The classic optical illusions shown in Figure 1 are caused when the brain invests too much confidence in predictions based on its rules of thumb, giving insufficient credence to raw sensory information. In the Café Wall (A), Kanisza Square (B) and Poggendorff (C) illusions, the context fools the brain into applying rules that lead us to see, respectively, parallel lines as converging or diverging; a complete square; and two segments of a diagonal line as offset rather than aligned. In the Rubin's Vase illusion (D) the brain vacillates between two alternative interpretations: a vase or two faces.

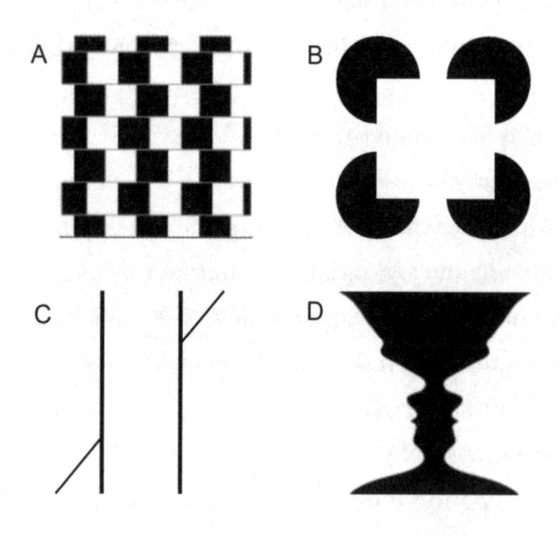


Figure 1: *These classic illusions reveal how the brain can slip up when it applies 'rules of thumb' to ambiguous sensory data*

Even if we are perceiving something for the very first time, the brain will always reach for what it already knows. To a metaphorical 'fish out of water' – a creature in an unfamiliar environment – a tree might look like seaweed waving in an ocean current until, through the interplay of its predictions and sensory prediction errors, the fish has updated its model to predict something more rigidly tree-like.

Clearly there's more to conscious perception than a stream of predictions flowing in one direction and a stream of prediction errors coming back in the other. Without some way to filter or vet the barrage of prediction errors, the mind would be quickly overwhelmed or easily misled. What altered states reveal is the importance of the brain's chemical tuning systems. Karl Friston, a neuroscientist based at University College London who helped pioneer the theory of prediction error processing, told me how these might work. For a start, he said, our brains don't need to pay any attention to sense data that exactly match their predictions:

> The breaking news is the surprising stuff – the bits you *didn't* predict. But you've still got to select which channel to tune into. Do you go for Sky News, BBC News or Fake News? You have to tune in and boost those prediction errors you think are going to have reliable, precise information, and ignore or turn down the volume of Fake News.

This is the role of attention: bringing the mind's processing power to bear on information that is not only newsworthy but also comparatively reliable. It's how the brain updates its models of the world to better reflect reality.[10] What appears to happen during all kinds of altered states of consciousness, and in mental illnesses, is that either too much credence is given to weak, unreliable news stories – the sensory 'noise' – or our models have become so entrenched they continue to be upheld in the face of contradictory evidence. This manifests, for example, in the repetitive, negative thinking of anxiety and depression, or persecutory delusions in schizophrenia. Holding rigid cognitive models or putting too much trust in weak evidence leads us to make false inferences about the world: the conspiracy theories, paranoia and outright hallucinations they peddle on the Fake News channel. Perhaps the auditory hallucination I experienced during my ayahuasca ceremony could be interpreted as my brain mistaking the dim echo of the rhythmically shaken chakapa, bouncing off the pillar behind my head, as the original source of that sound? A weak, unreliable signal that would normally be filtered from ordinary consciousness was instead given credence and magnified to create a very distinct perception. My mind was uncritically tuning into the Fake News channel.

Turning the volume up or down on competing streams of information is the job of 'neuromodulators', signalling molecules such as serotonin, acetylcholine, dopamine and oxytocin that influence or modulate the amount of communication between neighbouring nerve cells by binding to receptors in the membranes of their synapses. Neuromodulators mediate a host of brain functions, depending on the type and location of their target nerve cells, including mood (serotonin), learning (acetylcholine), reward (dopamine) and social bonding (oxytocin). Their levels also fluctuate cyclically over a roughly twenty-four-hour period, governing, among other things, the alternating conscious states of the sleep–wake cycle, which I explore further in Chapter 2. It's as if during the day the brain tunes into (mostly) dependable sources of information, such as Euronews, BBC News or CNN, and switches to Fake News at night. We now know that the classic psychedelics DMT, psilocybin and LSD achieve their dramatic shifts in consciousness – which have a lot in common with dreaming sleep – by binding to serotonin 2A receptors in the innermost layer of the cortex and temporarily jamming the brain with Fake News.

The drugs prescribed by doctors and psychiatrists to treat mental illness also work by twiddling our neuromodulatory knobs. By raising levels of serotonin, for example, antidepressants lessen the ruminative, negative thought patterns of depression, while anti-psychotics counter the delusions and hallucinations of psychosis by reducing the influence of dopamine.

The surprising truth that studies of sleep and altered states of consciousness reveal is that sometimes you need to go a little crazy in order to stay sane. To carry out the synaptic housework that restores flexibility to the brain's processing hierarchies and underpins their ability to find the most newsworthy, reliable information, they must be taken offline. And while our brains are unplugged, dreams, hallucinations and delusions inevitably arise. 'But if you want to relearn,' Friston told me, 'if you want to explore new hypotheses, you need to do this because these are exactly the same physiological, synaptic mechanisms that are necessary for neural plasticity and restructuring.'

If you are struggling to come to terms with your own mortality, or have lost any sense of direction in your life, new hypotheses of selfhood and your place in the world may be just what you need. Similarly, if you're stuck in the ruts of depression, addiction or obsession, your brain will welcome the opportunity to restore some plasticity to its unhealthily rigid behavioural and cognitive models.

Over the centuries humans have chanced upon numerous ways to oil the wheels of their minds. They have not only reconfigured their brains with hallucinogens but also by practising exacting spiritual disciplines. They have used the extreme physiological stresses of starvation, pain and discomfort to supercharge their serotonin levels, and they have subjected themselves to the rigours of sensory deprivation and isolation to overthrow their brains' information-processing hierarchies.

If all this sounds a little superhuman – too arduous for the average, untrained person to consider – there are alternatives so subtle you'll hardly know they're happening. It turns out that the mind's connections with reality can be broken relatively easily, without resorting either to chemicals or painful ascetic practices. In the trance state induced by a drug-free shamanic ritual, for example, all it takes to convince someone that their soul has been transported into an animal, another person or the spirit realm is the power of suggestion. This is a power that spans ancient and modern cultures. After a little gentle induction at the start of a show, a stage hypnotist can persuade suggestible volunteers that they have become chimpanzees or acquired X-ray vision. More productively, clinical hypnosis can help people overcome phobias and addictions, feel less pain in the dentist's chair or as a result of chronic conditions such as irritable bowel syndrome, and improve the efficacy of talking therapies for mental illness.[11]

We don't tend to think of it as hypnosis, but suggestion and expectation maintain a constant, hidden but powerful grip on our mood, behaviour and even physiology. The evidence for this is found in medicine. Doctors have long recognized that placebos – sugar pills and sham medical procedures – can have almost as big a clinical effect as active treatments, influencing not just mental well-being but also physical health. With their potions and rituals, faith healers down the centuries have exploited the very same top-down cognitive mechanisms: the hidden hand of suggestion and expectation. The efficacy of proven medical treatments can also be boosted by the right packaging for the pills or lotions, and how well our doctor 'sells' them to us as a cure for whatever ails us.

Is it any wonder that the course of a psychedelic trip is so powerfully influenced by 'set and setting' – a person's mindset and their surroundings? The drugs are known to increase suggestibility, ratcheting up the influence of ambiguous sensory data and unusual cognitive inferences. This could heighten anxiety and cause frightening hallucinations in a dimly lit, crowded nightclub, or it could enhance insight and feelings of connectedness in a quiet, cosy setting among friends.

Some very determined individuals have even learned how to master their own physiology at will. In the eighties, doctors at Harvard Medical School confirmed that some Tibetan monks can increase their body temperature using a meditation technique known as *g Tum-mo* yoga.[12] There's nothing inherently spiritual or mystical about the monks' ability. The endurance swimmer Lewis Pugh does much the same when he raises his core temperature before plunging into freezing water that would otherwise send his body into shock. In the summer of 2007, dressed only in swimming trunks, a cap and goggles, he became the first person to complete a long-distance swim in a crack in the sea ice at the North Pole.[13]

Feats like these – not to mention extraordinary states of consciousness such as meditation, trance and the psychedelic state – provide a bracing reminder that we never really experience the inner and outer worlds directly. Tinker with our neuromodulators, our sensory inputs or cognitive models and there is seemingly no end to the illusions, deceptions and physiological changes that can follow. Despite the faith we place in our thoroughly modern, rational minds, ordinary consciousness is far from immune to such distortions, though the effects are more subtle and easily overlooked. They only become apparent under special circumstances, for example through optical illusions, the placebo effect in a clinical trial, or through the bizarre tricks played by a stage hypnotist.

Intuitively, we believe the flow of information through our minds to be in one direction, from the bottom up, finding its way from the world into conscious awareness via our senses, when in fact most of the traffic flows in exactly the opposite direction. The brain's rough-and-ready models of reality, established over a lifetime of experience, infer or prejudge the causes of the body's sensory inputs, and it is these inferences, not raw sensory data, that are the stuff of conscious experience and behaviour. These models usually serve us perfectly well, but it is sobering to realize that prejudgement is hardwired into our brains, affecting not just perception but pretty much everything that goes on between our ears, what we do, and a lot of what happens inside our bodies.

Thankfully, we are not wholly at the mercy of prejudice. Perhaps uniquely among all the creatures on earth, we humans have the ability to reflect upon our thoughts and behaviour as they unfold from moment to moment – to observe ourselves and our inbuilt inferences with a certain amount of objectivity. Psychologists call it metacognition, meta-awareness or mindfulness. The good news is that this ability to see through the fog of delusion can be cultivated, and that certain altered states of consciousness may be particularly useful in this endeavour. The best evidence that this is possible even in the most delusional circumstances is lucid dreaming, which I explore in the next chapter. Even in the midst of a dream, we can recognize that our brains are deceiving us.

# 2

# *Dream On*

In a record shop in downtown Detroit, a radio DJ was raising money for children with muscular dystrophy. For the next nine days, the charismatic twenty-seven-year-old would be the star attraction at the store, spinning discs, entertaining onlookers with his spiel, shooting pool with friends and volunteers and being interviewed by local media.

The fundraiser got off to a cracking start – customers were enthusiastic and generous – but on the third day the DJ unexpectedly flew off the handle and, pulling on his coat, shouted: 'You're all trying to make a fool out of me! I'm quitting!' To everyone's relief, he calmed down and agreed to carry on, but three days after this outburst his behaviour took another nosedive. He became hostile towards female supporters and rudely refused a bowl of soup one had specially prepared, accusing her of 'mollycoddling' him. In the evening, after wolfing down a huge dinner, his mood turned euphoric and grandiose. He declared that he would take on 'every DJ in the US' in a competitive fundraiser in the window of a famous Detroit department store.

Overnight the euphoria ebbed away and in the morning everyone agreed he looked exhausted. He complained of pain in his joints and was seeing things: 'a grey mist hanging over the pool table like a spider web', a 'blue flame or luminescence' surrounding a young woman who brought his morning coffee. He fled a room in terror when he hallucinated flames spurting from the walls. Increasingly irascible, later that day while playing pool he tried to throttle his partner. Again he ranted that everyone was trying to make a fool out of him. They were talking behind his back! He wouldn't have it!

The poor man now had trouble focusing on the simplest of tasks. He was easily distracted by music or voices coming from other rooms and in the final days of the fundraiser the hallucinations were coming thick and fast. He reported feeling as though there were a tight band around his head that was gradually slipping down over his eyes, obscuring his vision. One moment he thought he was floating on a black cloud surrounded by the indistinct forms of ballet dancers, the next he was broadcasting live from the scene of a fire somewhere else in the city.

Grandiosity, paranoia, hostility, hallucinations, confusion – all the classic symptoms were there. Called upon to assess the DJ's mental health, a psychiatrist would have no trouble diagnosing the onset of schizophrenia. Unless, that is, he or she was first made aware of the true nature of the fundraiser: a 'wakeathon' popular in the mid-twentieth century among American DJs, especially those trying to revive flagging careers. It was the fifties and this particular fellow had recently been sacked from a minor radio station. He was now aiming to beat the previous record by staying awake for 220 hours straight. His attempt was closely monitored by local doctors and scientists from the Lafayette Clinic and Wayne State University, who in 1960 published a fascinating paper about their findings in the journal *Psychosomatic Medicine*, without disclosing their subject's identity.[1] The study added to growing scientific evidence that staying awake for five days or more unfailingly triggers the symptoms of psychosis.

By the ninth day of his record bid, the hapless DJ could barely speak or walk. Shortly after he reached his target 220 hours without sleep, he collapsed and was rushed to hospital. He slept for fourteen hours straight and awoke seemingly back to his former self, no worse for the experience. In the months and years that followed, others would repeatedly break the record for staying awake, most famously Randy Gardner, a student who in 1964 went for 264 hours without sleep. Sensibly, *Guinness World Records* no longer lists wakeathons, recognizing that to go without sleep for days on end is too risky. They have been declared illegal in many countries. Hallucinations, paranoia and violent mood swings are inevitable, even for those who start out in perfectly sound mental health. There are also possible long-term physical effects such as an increased risk of cancer and heart disease.

Going without sleep for just twenty-four hours – 'pulling an all-nighter' as students cramming for an exam call it – provokes mild psychotic symptoms. A study published in 2014 found that twenty-four healthy volunteers who stayed up all night experienced perceptual distortions, confused thinking and 'anhedonia' (lack of enjoyment of normally pleasurable activities) the next day.[2] The sleepless night also messed with their ability to filter out distracting but benign environmental stimuli, as revealed in the lab by a reduced capacity to moderate their startle response to sudden noises. Put another way, after a night without sleep subjects became less able to turn down the volume of the 'fake news' in their brains – the false inferences that something or somebody was threatening them. They were jumpy and a little paranoid. Tellingly, one of the scientific measures that pharmacologists use to judge the efficacy of a new antipsychotic drug is how well it restores this function of filtering out distracting but harmless auditory stimuli (known as 'pre-pulse inhibition').

Scientific studies involving more than forty-eight hours' sleep deprivation are now considered unethical, but a review of older studies published in 2018 found that complex hallucinations and disordered thinking began after two nights without sleep and delusions after three nights. After five nights the symptoms were indistinguishable from acute psychosis and subjects were incapable of differentiating between reality and delusion.[3]

Among patients diagnosed with schizophrenia or bipolar disorder, psychotic episodes are often preceded by prolonged insomnia. Psychosis and lack of sleep clearly go hand in hand, but what exactly is it about not getting enough that makes us go a little crazy? More to the point, how come sleep is so good at *stopping* us going crazy? The sleep stage most reliably associated with vivid dreams, known as REM (rapid eye movement) sleep, has a crucial role to play. Far from being mindless fantasies with no evolutionary purpose, dreams are vital for keeping us on an even emotional keel. The sleep scientist Matthew Walker and his fellow researchers at the University of California, Berkeley, have found that REM sleep tones down emotional responses to distressing stimuli experienced the previous day, reducing activity in a pair of almond-shaped structures deep inside the brain known as the amygdalae that generate positive and negative emotions in response to particular events.[4]

Combat veterans with post-traumatic stress disorder (PTSD) have disrupted REM sleep, which Walker thinks may explain why they continue to be plagued by recurrent nightmares. It's as if their brains are repeatedly trying and failing to heal traumatic memories. Remarkably, a drug called prazosin, which lowers blood pressure by reducing levels of the brain hormone noradrenaline, can improve the quality of REM sleep in people with PTSD and in the process liberate them from their appalling nightmares.[5,6] Noradrenaline tends to suppress REM. It is one of several 'neuromodulators' that determine our state of consciousness by controlling the flow of information through the brain. (I'll have more to say about how these molecular gatekeepers work later in this chapter.)

Even among people who have slept well the previous night, the need for emotion-regulating, dreaming sleep becomes ever more urgent as the day progresses. In Walker's sleep lab, they have discovered that healthy subjects rate photographs of facial expressions as increasingly angry, fearful or less happy the more hours they have been awake. This supports the intuitive notion that we judge other people's mood as increasingly negative and threatening the longer we've gone without sleep. More unexpectedly, the second part of the experiment found that volunteers who were allowed to take a short nap recovered much of their ability to make accurate judgements about others' moods, but only if they managed to get some refreshing REMs while they snoozed.[7] As Walker wrote in his book *Why We Sleep*: 'REM sleep is what stands between rationality and insanity.'[8]

Dreams themselves have a lot in common with psychosis. When you are dreaming you see and hear things that aren't really there and believe things that couldn't possibly be true. You are disoriented, forgetful and experience powerful emotions such as anxiety, elation and anger. To maintain our sanity during the day it seems we have to go a little crazy at night. This may seem counterintuitive, but all kinds of maintenance involve a complete suspension of normal operations. You can't fix the potholes in a motorway without temporarily closing a lane to traffic. An Internet bank can't overhaul its computer system without taking it offline for a few hours. And what better time to carry out essential maintenance than at night, when usage is light?

Dreams may reflect the brain's need to streamline or optimize the virtual-reality models that it uses to predict what's happening out there in the world.[9] In effect, everything we see, hear, smell, taste and touch is a prediction generated by these models. During the day, we refine the models by interacting with other people and everything else in our environment. We update them – we learn – by resolving 'prediction errors', the discrepancies between the models' predictions and data streaming from our senses.

At the neural level, this entails creating and strengthening the synapses that connect nerves. This is what neuroscientists mean when they say 'nerves that fire together, wire together'. These increases in connectivity are the basis of 'associative learning' when two environmental stimuli occur almost simultaneously, suggesting there is a high probability of a cause-and-effect relationship between them. The trouble is that, while daytime learning makes the brain's models increasingly *accurate*, all those additional, stronger connections render them increasingly *inefficient*. Profuse connections not only take up limited space and consume valuable energy, they also make everything unnecessarily complicated and overly specific. The upshot is that the models aren't much use in novel situations.

The brain's predicament is rather like a large, established business that has got itself tied up in red tape, losing its ability to adapt quickly to new market conditions. Over time redundant systems for dealing with very specific scenarios encountered in the past, that are unlikely ever to crop up again, have accumulated. As a result, the business is now bogged down in time-consuming, distracting minutiae and has lost its ability to adapt quickly to new circumstances.

Multinationals and brains need a way to cut through all this red tape without jeopardizing their operations. To understand the solution that evolution appears to have come up with, it's worth remembering that the brain's models are a lot like scientific hypotheses. Both are predictions about how the world works based on limited evidence, and whenever a scientist (or your brain) puts forward a hypothesis, the idea can then be kicked around and tested further against new evidence as it becomes available. But at the outset there will be any number of competing hypotheses that also fit the evidence and only limited resources to investigate them. How can brains or scientists predict which is worth pursuing?

When trying to decide between hypotheses, scientists often invoke an ancient principle known as 'Occam's razor', attributed to a fourteenth-century Franciscan friar called William of Ockham. Essentially this states that they should plump for the *simplest* idea – the one that makes the fewest assumptions. My favourite formulation of Occam's razor is: 'When you hear hoofbeats, think horses not zebras.' Say you were somewhere in Surrey (where William of Ockham was born) when you heard the thunder of hooves, you'd be a fool to infer that a herd of zebras had escaped from a local zoo and were galloping your way – even if there was a minuscule possibility this might actually be true.

Sleep may deploy the same simplifying principle. The idea is that in sleep, your brain wields Occam's razor to prune the profuse neuronal connections that have sprouted during the day's learning experiences. This has the effect of making its models simpler and more efficient. The neuroscientists Giulio Tononi and Chiara Cirelli from the University of Wisconsin-Madison have proposed that overnight the brain saves energy and restores flexibility by indiscriminately downgrading the strength of all its synapses by a similar amount. As a result, the strongest connections get a little less strong – introducing some 'slack' into the models that will allow for future fine-tuning – while the weakest disappear completely.

This admirably simple but effective strategy, which Tononi and Cirelli call 'synaptic homeostasis', prevents the brain's models of the world from getting too complex and hidebound.[10] It minimizes the number of speculative premises they draw upon, in much the same way a scientist rejects hypotheses that depend on a lot of assumptions – each supported only by weak evidence – in favour of a simpler hypothesis. In other words, while you sleep, synaptic homeostasis throws out the kind of highly speculative models that depend upon things like zebras escaping from zoos, while preserving a simpler, more plausible explanation for the sound of distant hoofbeats, namely galloping horses.

Like a washing machine going through a preset programme of washing, rinsing and spinning, the brain may go through a fixed programme of synaptic pruning while we sleep, addressing different kinds of memory in turn. When researchers monitor the electrical activity of sleeping volunteers' brains, via electrodes glued to their scalps, they see clearly defined cycles of activity lasting about ninety minutes each. Within each cycle there are three stages of increasingly deep, non-REM sleep, followed by a single period of REM sleep when the brain becomes almost as active as it is during the day. Dreaming can occur at any stage of the cycle, but vivid dreams with highly emotional, narrative elements and lots of action occur exclusively during REM. It makes sense, therefore, that scientists have found REM sleep to be strongly associated with emotion regulation and the learning and relearning of motor skills. Maintenance work – synaptic pruning – involving emotional or behavioural models can only be carried out safely during the paralysis of REM sleep when our skeletal muscles (apart from those controlling eye movements) are temporarily disabled, preventing us from acting out our psychotic dreams.

When people are deprived of REM sleep this seems to inhibit

their ability to recover from a frightening experience. In an experiment in 2011 at the Max Planck Institute of Psychiatry in Munich, volunteers were conditioned to experience fear on cue: whenever they saw a particular visual stimulus they received a mild electric shock, until eventually the stimulus itself provoked fear. In subjects who were prevented from dreaming for the next few nights, despite repeated training sessions in which the stimulus was never followed by a shock, their fear response persisted. By contrast, in people allowed to dream, the stimulus no longer evoked fear. For them, it was as if the emotional slate had been wiped clean as they slept.[11]

What wakeathons and sleep deprivation studies reveal is that our dreams train us to be healthy, functioning humans. Go without REM sleep even for a night and your capacity to regulate your emotional life starts to disintegrate. Go without sleep for days on end, like our hapless fifties DJ, and you will inevitably become confused, paranoid and prone to emotional outbursts and hallucinations. Your body's homeostatic control mechanisms, which keep things like appetite and body temperature at optimal levels, will also be compromised. Perhaps as a result, getting fewer than seven hours' sleep a night is not only associated with poor mental health but also impaired physical health, with increased risk of cancer, diabetes, heart disease and Alzheimer's. Sleep deprivation experiments with rats suggest that if this dysregulation goes on for long enough it eventually results in death.[12]

Without sleep, none of your predictive models about how the world and your body work will be safe, even the 'sensorimotor' memory of walking. They all have to be *optimized* while you sleep. When I asked the renowned sleep researcher Allan Hobson why we dream, he explained:

> When someone reads the word 'memory' in your book they won't be thinking about things like remembering how to walk because they assume you just *know* how to do that. How do you *do* things? How do you hold a cricket bat? All these things are taken totally for granted, as if they were built in. They're not built in. They need to be learned and relearned. Memory is fleeting and if it weren't renewed it would probably disappear. Even the sense of self could be considered remembered and has to be relearned every night.

After a long and distinguished career in sleep research – at the time of writing he is well into his ninth decade – the emeritus professor of psychiatry at Harvard Medical School now spends much of his time writing books on his farm in Vermont and curating his very own Museum of Dreams. In recent years he has worked closely with Karl Friston at University College London and Charles Hong at Johns Hopkins University in Baltimore, US, to develop the hypothesis that our virtual-reality models of our bodies and the outside world are optimized as we dream.

Their idea builds upon a theory that is gaining increasing acceptance among neuroscientists: that the brain is an 'inference engine' whose job is to predict the hidden causes of its sensory inputs. The engine draws on past experiences to predict or 'model' what's happening in the world, refining its models using inputs from the five senses. According to the theory of prediction error processing, it does this by exchanging messages between consecutive layers in a cortical data-crunching hierarchy that stretches from the senses to the prefrontal cortex (see Figure 2). Each layer attempts to predict sensory inputs from the one immediately below, with any mismatches or 'prediction errors' passing upwards through the hierarchy, updating the model. A flashbulb moment of conscious perception or recognition occurs when errors have been minimized throughout the hierarchy.

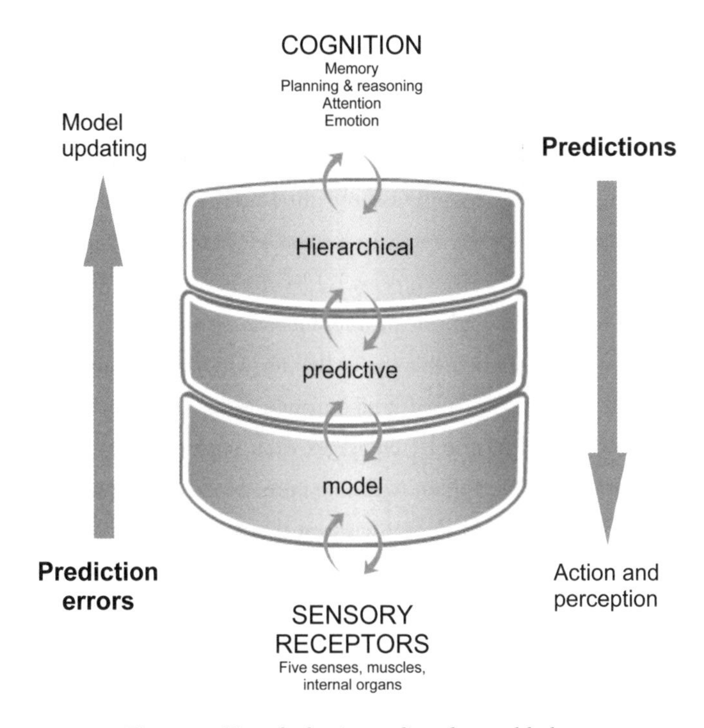


Figure 2: *How the brain predicts the world: three layers of a prediction error processing hierarchy*

Hobson first proposed more than forty years ago that as we lose consciousness and slip into non-REM sleep it's as if outer gates regulating the flow of information through the brain are swinging shut, blocking nearly all input from the senses and deactivating entire brain regions. Animal studies suggest these 'gates' are located at the top of the spinal cord in the brainstem, where bundles of nerves regulate vital bodily functions such as breathing and heartbeat, and consciousness itself. Later in the night, as we enter

REM sleep, conscious activity is allowed to return in the form of dreams. The gateway from the senses remains shut, however, and now the gate to the skeletal muscles also swings closed, causing sleep paralysis. The gatekeepers in the brainstem responsible for determining brain activation – whether we are awake, deep in non-REM sleep or in REM sleep – are the neuromodulators serotonin, dopamine, noradrenaline and acetylcholine. Their levels fluctuate across a roughly twenty-four-hour cycle, opening and closing the gates of the mind in a strict sequence. Serotonin and noradrenaline appear to suppress REM sleep, for example, while acetylcholine and dopamine promote it.

Hobson has dubbed this the 'AIM' model of consciousness: A for brain Activation; I for Input-output gating; and M for Modulation.[13] The AIM model envisages each state of consciousness occupying a discrete region within a three-dimensional space. The state we find ourselves in at any given moment – awake, in non-REM sleep or REM sleep, for example – is determined by how energetically active our brains are (Activation); whether sensory input and motor output circuits have been switched on or off (Input-output gating); and the relative amounts of neuromodulators being synthesized in the brainstem (Modulation). Activation can be measured using brain-scanning techniques such as MRI; Input-output by measuring responsiveness to sensory inputs and electrical activity in muscle tissue; and Modulation by measuring neuromodulator levels in the brainstem (currently only possible in lab animals). Just as there are an infinite number of points in three-dimensional space, Hobson believes infinite variations of consciousness are possible, depending on the relative levels of A, I and M. Nonetheless, a healthy brain usually settles into one of three states: wakefulness, non-REM sleep or REM sleep.

Things start to go awry when the gates regulating the flow of information through the brain aren't opened and closed to the correct extent and in the appropriate sequence. We have already seen that persistently high noradrenaline in the brains of sleeping PTSD patients interferes with their REMs, preventing the resolution of traumatic memories, and that a blood-pressure drug that reduces levels of noradrenaline can help. By contrast, the problem for people prone to psychosis appears to lie with the dopamine gatekeeper.

More than 200 years ago, philosophers noticed that the delusions and hallucinations of mental illness resemble the delusions and hallucinations of dreams. The German philosophers Immanuel Kant and Arthur Schopenhauer observed that 'the madman is a waking dreamer' and 'dreams are short madness and madness a long dream'. In the nineteenth century, the British neurologist John Hughlings Jackson predicted: 'Find out all about dreams and you will find out all about insanity.'[14] They were all on to something important. In schizophrenia and the manic phase of bipolar disorder, it's as if REM sleep is intruding into waking consciousness, causing psychotic symptoms such as grandiosity, seeing or hearing things that aren't really there, and paranoia. By reducing the availability of dopamine in the brain, antipsychotic drugs alleviate these symptoms by nudging this particular gate towards the closed position. Tellingly, one of the side effects of antipsychotics is reduced REM sleep.[15] Conversely, in previously healthy people experiencing chronic sleep deprivation, the dopamine gate may swing wide open as the need for REMs becomes increasingly urgent, triggering the dream-like waking hallucinations and paranoia that once plagued students and DJs attempting to break the Guinness World Record for staying awake.

Psychosis has an intimate relationship with all kinds of altered states – from trance to meditation and the psychedelic state – a relationship I'll explore further in later chapters. As humans we find ourselves in a strange predicament. If, as seems increasingly likely, the brain's modus operandi is to generate virtual realities, how can we ever be sure that what we are experiencing is real and not just another dreamy fantasy? While we are awake our senses report back prediction errors that constrain the brain's infinite creativity, moulding what we see, hear, taste, feel and touch, keeping us grounded in reality as far as that is possible, but in truth we never experience the world directly. We can only infer what's happening.

If that were the end of the story it would be somewhat depressing. We'd be little better than conscious automata, aware but devoid of critical self-awareness. Thankfully we have been spared this fate, but in a way entirely consistent with prediction error processing. After all, if an organism can model the world around it, including the thoughts, feelings and intentions of its fellows, why shouldn't it also be able to model *itself*, given the requisite neural upgrade: an extra layer in its processing hierarchy? The ability that *Homo sapiens* has evolved that other conscious creatures seem to lack is metacognition, which is the capacity to think about thinking.

Metacognition comes with an impressive suite of new functionalities, including insight, planning and objectivity – skills that Hobson has dubbed 'secondary consciousness'. The more primitive, 'primary consciousness' seen in other animals and human infants restricts them to first-person, present-centred emotion and perception – in other words the automatic stuff that dreams are made of. As children mature, unlike other young primates they start to develop the ability to observe their own thoughts: to split or 'dissociate' their consciousness into an actor and an observer.

If all goes well, secondary consciousness broadens and deepens as we grow older and wiser. But it melts away in our dreams. Like muscle paralysis, this may be vital before overnight maintenance work on intermediate levels of the processing hierarchy can begin. However, there is a rare, evanescent dreaming state that has features of both primary *and* secondary consciousness. In this state people become aware that they are dreaming and that everything they are experiencing is a virtual reality story cooked up by their sleeping brain. It's called lucid dreaming and is proving wonderfully useful for scientists trying to solve the riddle of how the brain creates secondary consciousness.

'Lucid dreaming is a superb test case because it's a hybrid state,' said Hobson. 'You *know* you're sleeping, you *know* you're not awake. So part of your brain, in a way, is awake, and part is asleep.' By imaging the brain of someone having a lucid dream and then comparing the images with non-lucid REM sleep, researchers can pinpoint the regions essential for metacognition and secondary consciousness.

Many of us remember having had lucid dreams in childhood (their frequency peaks at six or seven years of age). The first scholarly account of the phenomenon was written more than 150 years ago by the French aristocrat d'Hervey de Saint-Denys, but scientists remained deeply sceptical about the whole thing until comparatively recently. Are people really asleep when they think they're having a lucid dream? And even if they are, how can an independent observer investigate a subjective experience that happens while the person is paralyzed from head to foot and incapable of communicating? In the seventies, a psychology student at Hull University in the UK called Keith Hearne hit upon a brilliant solution to these problems that had somehow eluded everybody else. He reasoned that since the only skeletal muscles that remain active during REM sleep are those moving the eyes, someone who becomes lucid in a dream and gains some control over the experience should be able to use their eyes to communicate this. On the morning of 5 April 1975, with a prearranged sequence of seven left–right eye movements, a subject in the university's sleep lab became the first human in history to send a message to the waking world from inside a dream.[16]

The validity of the technique was confirmed by Stephen LaBerge at Stanford University in 1980, proving beyond all reasonable doubt that lucid dreaming is for real. Nonetheless, it would be many years before neuroscientists viewed lucid dreams as anything more than a way to do fun things like flying and acting out erotic fantasies. So when Hobson invited a psychologist at Goethe University in Frankfurt called Ursula Voss to collaborate with him on a study of lucid dreaming she was horrified. At a public lecture in 2014, she laughed at the recollection: 'I thought, "Oh no! Let's do something more *scientific*."' Hobson was the 'dream pope', she said – his reputation preceded him – but surely lucid dreaming was no more than a self-indulgent game for people who didn't have to get up very early in the morning?

Fortunately Hobson persisted, Voss relented, and the result was a groundbreaking study published in 2009 in which they used electroencephalography (EEG) to show that the front of the brain, which is usually dormant in dreams, awakens during lucid dreaming. Six psychology undergraduates who claimed to have frequent lucid dreams were trained to signal when they were dreaming lucidly with a pre-agreed sequence of eye movements, then spent several nights in the university's sleep lab with electrodes glued to strategic locations on their scalps recording the telltale electrical oscillations of brain activity. Later, when the researchers compared traces during lucid and non-lucid REM sleep, they discovered a surge in high-frequency, 'gamma' waves at the front of the students' brains whenever they were having a lucid dream, peaking at 40 Hz – a frequency usually only seen in waking consciousness. Remarkably, even though the students were fast asleep, the signal was almost as strong as during waking.

Three frontal areas of the brain closely associated with metacognition – the dorsolateral prefrontal cortex, orbitofrontal cortex and precuneus – are now known to come back online during lucid dreams (see Figure 3).[17–19] People who frequently experience them have more grey matter and higher waking activity in these parts, suggesting a more metacognitive mindset.[20] When someone becomes lucid during a dream, his or her brain switches from the purely emotional state of primary consciousness to the distinctively human self-awareness and volition that characterize secondary consciousness. At the front of the brain, the dorsolateral prefrontal cortex in the right hemisphere, and the orbitofrontal cortex in both hemispheres, reawaken. The former is involved in planning, decision-making and working memory, while the latter is associated with learning and emotional regulation. The region most strongly activated is the precuneus, implicated in our sense of agency and ability to reflect on ourselves. Between them, these parts of the brain allow us to 'dissociate' from what's happening to ask the question, 'Am I dreaming?'

Intriguingly, there is a striking overlap between lucid dreaming hotspots and the regions that have expanded most during our recent evolution, distinguishing our brains from those of our primate relatives. What's more, the very same regions – intimately involved in metacognitive abilities such as self-reflection, insight, working memory and decision-making – are dysfunctional in people with mental illnesses characterized by psychosis.[21]

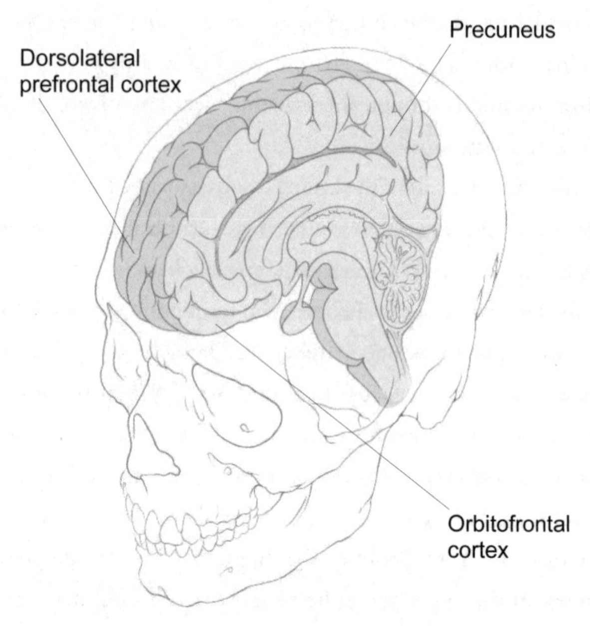


Figure 3: *Lucid dreaming hotspots in the right hemisphere of the brain. All are closely involved in metacognition*

Through the study of dreams a picture is starting to emerge of a recently evolved form of consciousness – metacognition or secondary consciousness – that is unique to humans, is regularly taken offline for maintenance during sleep and is faulty in people going through a psychotic episode. 'An animal needs to have a highly developed mind in order to go out of it,' quipped Hobson and Voss in 2011 after reviewing all the scientific evidence to date.[22] Happily for *Homo sapiens*, there is also ample evidence that the lucidity of secondary consciousness can be cultivated. Of course it helps to fit as much REM-rich sleep as possible into our busy schedules. And by following some simple guidelines (see pages 57–65), with persistence one can learn how to dream lucidly.

It gets harder with increasing age, but a few years ago I began to have lucid dreams of my own, the first since childhood, at the age of fifty. Some people have cultivated the rare ability to become lucid in the midst of their dreams, which has happened to me on a few occasions, but much more common are wake-induced lucid dreams or 'WILDs', which take shape at the pleasantly woozy frontier between the waking and dream worlds. The trick is first to still the mind through a meditative practice, such as focusing on the breath, then maintaining an open-minded, non-judgemental awareness of any thoughts, images and narratives that pop into the mind as it drifts into sleep. If you manage to hang on to sufficient objective awareness to notice when your internal dialogue or storyline starts to get a little wacky or irrational, you're in. Of course, at this point you may wake yourself up or slide into unconsciousness, but with dedicated practice you can draw out those moments of lucidity in the midst of all the dream madness.

In principle, any technique used to become a lucid dreamer could have the knock-on benefit of boosting lucidity in the daytime. One proven way to foster greater lucidity – in both waking and sleeping consciousness – is meditation, which I investigate in the final two chapters of this book. More speculatively, the psychosis-like state of consciousness provoked by psychedelics may also have a role to play in boosting our metacognitive faculties. As already mentioned, one study has suggested that in the weeks after drinking ayahuasca subjects' scores on some measures of mindfulness are boosted. A more recent study conducted during a retreat at a monastery in the Swiss Alps (unpublished at the time of writing) found that the mindfulness scores of meditators were significantly increased in the days after they were given psilocybin, compared with those given a placebo.[23]

Meditation and psychedelics are age-old methods for fostering metacognitive well-being, but Voss and her colleagues appear to have invented a completely new technique. After their seminal study in 2009 demonstrating that frontal regions of the brain come back to life during lucid dreams, progress at Voss's lab essentially ground to a halt. They wanted to prove that the 40 Hz gamma waves they had detected at the front of the brain *caused* the associated shift in consciousness and were not simply a consequence of it. The problem was that, like netting a rare butterfly, catching lucid dreams in the lab was proving infuriatingly difficult. In their initial study, for example, of twenty students who claimed to experience frequent lucid dreams in the comfort of their own beds, only three succeeded in the unfamiliar environment of the sleep lab with the noise of equipment and with their heads bristling with electrodes. Each only managed one lucid dream in the lab.

Ungainly, head-mounted devices that emit light or sounds when the characteristic REMs of dreaming sleep are detected are commercially available. The idea is to prompt the dreamer to become lucid, but in the lab, if subjects managed to get any sleep at all, these devices usually just woke them up. Finally, after years of failed attempts to trigger lucidity using jerry-rigged variations of this equipment, Voss and her colleagues tried a piece of kit someone had bought online that silently and unobtrusively passes a weak alternating current through the scalp – so weak that the volunteers couldn't feel it. The tACS (transcranial alternating current stimulation) device was switched on a couple of minutes after REM sleep was detected. Subjects were then woken up and asked to rate how lucid their dream had been. In the course of the study, out of a wide range of frequencies applied (including a placebo, zero setting), only 40 Hz (and to a lesser extent 25 Hz) triggered lucidity, setting off brainwaves in the corresponding gamma EEG band.[24]

At a stroke, Voss's lab appears to have discovered not only a reliable way to evoke lucid dreams (a technology that dream entrepreneurs are no doubt already busy developing commercially for this purpose), but has also added weight to evidence from other scientists that forty electrical pulses per second is the magic frequency that the brain uses to orchestrate secondary consciousness: synchronizing activity in disparate regions. In the coming years, more research by other sleep labs will be needed to confirm these results, but Voss is already testing her findings in the clinical realm.

At a specialist clinic in Friedrichsdorf near Frankfurt, she is using tACS at 40 Hz to trigger lucid dreams in patients with PTSD as they sleep, helping them confront and heal the crippling anxieties that manifest in their nightmares. She is also pioneering a treatment for schizophrenia and obsessive compulsive disorder (OCD) that involves tACS sessions in awake patients to foster greater metacognitive awareness, with some promising, albeit preliminary, results. In one pilot study, major improvements were seen in seven patients with severe OCD who had failed to respond to drugs or cognitive behavioural therapy (CBT). Each patient underwent between eight and twenty treatment sessions spread over several weeks. One of them, a fifty-year-old man who had suffered from OCD since he was eleven, would often spend all day and night in the bathroom obsessively washing and trying to record his every thought in a diary. Shortly before the pilot study he had been hospitalized for eight months. Conventional treatments didn't help. After eight tACS sessions spread over two weeks, however, there were major improvements in his condition, allowing him to move to a therapeutic living community where he took a part-time job as a gardener. Voss and her colleagues report that he has been stable for the past three years.[25]

Preliminary tests of tACS in patients with schizophrenia have also been promising. Voss recalled one patient who regularly heard a voice inside his head telling him to jump out of windows or from rooftops, but after the treatment he learned to negotiate with the voice. He would refuse to put his life in danger, she said, 'So the voice would then say, "well at least jump down from the table..." The patient would come to my office, very proud, and say, "I jumped down from the table, but I didn't do anything else!"'

This neatly illustrates the adaptive 'dissociation' that becomes possible with metacognitive awareness, allowing us to adopt an independent, objective vantage point on whatever's happening, whether that's a voice in our head, obsessive thoughts or behaviours, or the fears that stalk our nightmares. In the next chapter, I will review evidence suggesting that even playing video games can help nurture this ability.

Unusually high-amplitude gamma waves (at 25–42 Hz) have been recorded in the brains of experienced meditators in Tibetan Buddhist traditions, not only while they practise a form of 'objectless' meditation but also to a lesser extent in ordinary, waking consciousness.[26] Combined with evidence from Voss's dream team and other researchers, this suggests that the lucidity of secondary consciousness – entrained by gamma-wave synchrony – can be nurtured. Metacognitive insight and self-awareness, surely the crown jewels of human consciousness, can be polished. We are not doomed to be swept along in a dream-like, hallucinatory state of mind, helpless in the face of our fears and compulsions. We may be dreaming, but we can awaken.

The paradox of altered states is that only by unplugging ourselves from ordinary sensory reality for a little while can our brains streamline or simplify their models to more closely reflect that reality. One can envisage a future society in which, armed with this precious knowledge, traditional methods for awakening the mind are augmented with hi-tech gadgetry to enhance secondary consciousness. Unlike the 'brainwashing' technologies of dystopian fiction and intelligence agencies' Cold War fantasies, these tools will be used to help dispel the nightmares, obsessions and delusions to which the human mind is prey. They will help future generations keep it real.

### *How to Dream Lucidly*

As a child, Mary Arnold-Forster discovered she could fly. Halfway up the dimly lit staircase that led to her nursery, a landing opened onto a conservatory where monstrous creatures lay in wait for her.

> In some of the first dreams that I can remember I was on that staircase, fearful of something which I was especially anxious never to have to see. It was then that the blessed discovery was made… it was just as easy to fly downstairs as to walk; that directly my feet left the ground the fear ceased – I was quite safe; and this discovery has altered the nature of my dreams ever since.

Arnold-Forster cultivated this new-found ability to change the narrative of her dreams until she could dream lucidly at will. Awake, she led the constrained life of an Edwardian lady. Asleep, she enjoyed total freedom. In her enchanting book *Studies in Dreams*, which she published in 1921 at the age of sixty, she wrote:

> Only in sleep the imagination is set at liberty and is free to exercise its fullest powers. Sleep which brings us our dreams fulfils the eternal need within us, the need of romance, the need of adventure; for sleep is the gate which lets us slip through into the enchanted country that lies beyond.[27]

Unfortunately it becomes increasingly difficult to enter this realm the older we get. In the maturing brains of children and adolescents, lucid awareness often arises spontaneously in dreaming sleep. By the age of sixteen more than half say they have experienced lucid dreams, but their incidence drops dramatically thereafter.[28] With persistence, however, adults can find their way back into this land of infinite possibility, and the first step simply involves saturating your waking mind with thoughts of lucid dreaming.

In the sixties, the 'dream pope' Allan Hobson was lent a copy of Arnold-Forster's book by the hostess at a dinner party. The scientist was sceptical about the whole thing, but reading the book in bed later that evening he fell under its spell. 'I had lucid dreams all over the place that night, and I didn't even try to induce them. They just happened because I'd read the book.' One in particular stuck in his mind:

> I was teaching at Harvard – where I did teach – and I took the students outside onto the grass to show them I could fly. I flew up above them about 30 feet high and continued to lecture as I was flying, and

then I came back down to earth and levitated, just to show them that was easy as well. I held myself still five feet off the ground and then dropped to my feet and said: 'You see, it's really very simple!'

In addition to reading and thinking about lucid dreams during the day, there are several other practices that are proven to increase their likelihood. They are designed first to improve dream frequency and recall, then identify telltale signs that you're dreaming, and finally trigger lucidity and control over what happens.

**Keep a dream journal**

Unless you record them, it will be as if your dreams never happened. The thrills, the strange encounters and revelations will cease to exist the moment you awaken. Dreams were Hobson's day job: he'd been recording his own nocturnal escapades for years. Keeping a journal of your dreams not only keeps the intention to dream lucidly to the fore of your waking mind, it also trains you to recall them and recognize when you're in the midst of one.

Physically moving when you awaken seems to wipe dreams from working memory, so the instant you realize you've been dreaming lie perfectly still and recall as much of what happened as possible. Turn it over in your mind for several minutes before reaching for the notepad and pen you will have left at your bedside and scrawling a few keywords that will allow you to recall the dream in the morning and record it in detail. I prefer this to switching on a light and writing out the whole thing straight away, which will fully awaken you and possibly your partner too.

### Identify dream signs

Within a week or two you may have enough material to start identifying 'dream signs': themes that pop up over and over again that can be used to alert you to the fact that you're dreaming. For some reason I often dream about climbing, though I rarely scale anything higher than a stepladder during the day. Cued by urgent signals from their bladders, lots of people dream about searching for a toilet. As a result of sleep paralysis, being unable to run or move is another common sign that you're dreaming. Others are all your teeth falling out or arriving at work only to discover you've forgotten to put your clothes on.

Situations in which you meet and interact with people you haven't seen for a very long time, the deceased or celebrities are also commonplace. In my first experience of lucid dreaming as an adult, I was being wheeled down a hospital corridor towards an operating theatre when the orderly pushing my trolley leaned in close and whispered: 'Your surgeon today will be Dawn French.' The immediate realization I was dreaming was so exciting I woke up, but it was progress.

**Perform status checks**

For months I carried a ragged file card around in my pocket which I pulled out and read whenever I heard a siren (which in London was a regular occurrence, day and night). Written on the card in thick green ink were the words 'Are you dreaming?' This is known as a 'status check'. The idea is that if you get into the habit of questioning reality repeatedly during the day, you'll carry the habit into your dreams. You could ask yourself if you're dreaming whenever you enter or leave a room, or hear music in your head, for example. Some people go further and perform actions to test reality, such as jumping in the air (in a dream it will take longer than it should to come back down to earth), trying to push a finger through their palm, examining a clock face or reading text (in a dream the numbers or words may be jumbled and illegible).

I once tried to convince a friend we were both dreaming by suggesting he jump up and down. When that didn't work (I wasn't sufficiently lucid to realize *he* was unreal) I reclined in mid-air as if floating on my back in a swimming pool, which certainly did the trick. In another dream, in which I found myself in a bar, I managed to convince both myself and my drinking companions that none of it was real by pushing my hand through one of their faces – though I wouldn't recommend this as a regular daytime status check.

**Maximize your REMs**

Alcohol suppresses REM (rapid eye movement) dreaming sleep, so you should abstain on nights you want to dream lucidly. Ideally you'll also allow yourself the luxury of a lie-in, because most dreams occur during the REM periods towards the end of sleep, which are longer, lasting up to an hour (the first REMs of the night may last only ten minutes).

I have found supplements such as huperzine A and galantamine particularly useful for promoting lucidity. They work by inhibiting the enzyme that breaks down acetylcholine, a neurotransmitter that is active during waking hours, promoting alertness, attention, learning and memory. At night – operating in the absence of the neurotransmitters noradrenaline and serotonin – it produces REM sleep. Supplements that prevent the breakdown of acetylcholine not only increase the likelihood of dreams, they also boost their vividness and lucidity, and improve your chances of remembering them. They should only be taken at the recommended dose on an occasional basis, however, as they have been known to cause mild side effects including nausea, vomiting and diarrhoea. You shouldn't take them if you're pregnant or lactating.

Another way to boost dream recall and lucidity is to set your alarm to wake you up at a time likely to coincide with an REM period. This is particularly useful if you rarely awaken in the night and so remember few, if any, of your dreams. Setting your alarm for four-and-a-half hours after lights out – roughly three sleep cycles later – will stand a good chance of awakening you during the best REMs of the night. Research suggests that when the alarm goes off, getting up and reading or writing for thirty minutes or so before going back to bed is a particularly effective way to tap into this rich seam of dreams (see 'Night schedule', below).[29]

**Meditate**

'Only connect,' wrote Mary Arnold-Forster's uncle, the novelist E. M. Forster, of the need to break down the social barriers between people. In the context of lucid dreaming, 'only disconnect' might be more apposite. In a lucid dream there are two selves: the subjective self experiencing the fantasy, and an objective, observing self that sees through it. So whereas an ordinary dream is akin to the delusion of psychosis – a complete loss of touch with reality – to become lucid you must realize that what's happening is not real. This is a mild, healthy form of what psychologists call dissociation.

There are obvious parallels with meditation and mindfulness practice here, which also require the recognition that much of what's going on inside your head – the maelstrom of sensations, thoughts, emotions and feelings – only has a tenuous connection with objective reality. So it makes sense that regular mindfulness and meditation practice during waking hours have been found to increase the frequency of lucid dreaming.[30]

**Practise MILD**

The American dream researcher Stephen LaBerge has developed and tested a technique called 'mnemonic induction of lucid dreams' or MILD which you practise as you return to sleep after waking in the night. It relies on prospective memory: remembering to do something in the future. The basic steps are as follows:

- Recall a dream, ideally one from which you've just awakened.
- Identify a recurrent anomaly or dream sign within it, such as flying or paralysis (see above).
- Visualize returning to the dream and recognizing that this anomaly only happens in your dreams.
- As you do this, inwardly repeat something like: 'I'm going to recognize that it's only a dream.' Each time you say it, imagine carrying out a particular action, such as giving the thumbs-up sign or doing a fist pump. This will prove to yourself that you're in control of the dream content.
- Maintain a calm, meditative focus: whenever distracting thoughts or emotions arise, simply let them go without judging yourself.
- Repeat your visualization until you fall asleep.

**Night schedule**

LaBerge and his colleagues recently tested the following schedule on people taking part in a lucid-dreaming retreat. In a single night, 42 per cent of participants had a lucid dream after taking 8 milligrams of galantamine, compared with 14 per cent of those who took a placebo.[31] To avoid a build-up of sleep deprivation, he recommends only taking the supplement or using scheduled awakening once a week.

- Set an alarm to wake you after four-and-a-half hours, or longer if you usually have trouble falling asleep.
- As you turn out the light, repeat several times with conviction 'Tonight I will know when I'm dreaming.'
- When the alarm goes off, get out of bed straight away and take 8 milligrams of galantamine.
- A period of wakefulness will increase your chances of having a lucid dream. Stay out of bed for thirty minutes and keep yourself busy with a quiet, wakeful activity, such as reading about lucid dreaming or noting down a dream.
- Return to bed and practise MILD until you fall asleep.

# 3

# *Holidays from Reality*

When I played *Dungeons & Dragons* with my friends at an English boarding school in the late seventies, the game offered us a fantastical escape from the daily round of double maths, rugby, bullies and boredom. It transported us to subterranean realms where we were heroes with supernatural powers battling monstrous trolls for treasure.

Some forty years later, 'massively multiplayer online role-playing games' (MMORPGs) – a term coined by the American video-game developer Richard Garriott – offer a similar escape route for millions, myself included. The archetypal MMORPG, *World of Warcraft*, unfolds across the fictional continent of Azeroth and is a direct descendant of the *Dungeons & Dragons* game I enjoyed as a schoolboy – though back in the seventies its technological wizardry would have seemed more wondrous than anything my friends and I could have dreamed up. Both worlds are populated by warring orcs, dwarves, elves and wizards, but whereas its forerunner was played in private by some of the least cool kids in school using little more than dice and imagination, fantasy games like *World of Warcraft* are realized on computer screens in lurid colour and are mainstream entertainments played by adults and children alike.

Gaming is big business, easily eclipsing music and film sales.

In the UK, the market was worth £5 billion in 2017. Roughly 40 per cent of Britons between six and sixty-four years of age played a video game in 2016, each averaging nine hours' play per week. Global subscriber numbers for *World of Warcraft* have fallen in recent years as competitors enter the field, but in 2008 a staggering 10 million people – roughly the population of Sweden – were paid-up citizens of Azeroth. By 2013, subscribers had spent more than 6 million years between them in this parallel world.[1–4]

The allure of online role-playing games is that they provide a brief, blessed holiday from reality. They allow players to 'dissociate', escaping the stress and tedium of ordinary existence. It goes without saying that dissociation is a double-edged sword. Few would dispute that regular, temporary flights from reality, through books, films, TV and games, are a good thing. Highly absorbing, dissociative activities, whether they involve hobbies, work or sport, that provide challenges yet lie within our capabilities create the pleasurable state of focused consciousness known as 'flow'. But when psychiatrists talk about dissociation it is usually in the context of negative symptoms such as depersonalization (feeling detached from your body, thoughts and feelings) and derealization (a sense that everything is unreal), which are conceived as mental defence mechanisms against traumatic events that often occurred many years earlier, perhaps in childhood. Under these circumstances dissociation numbs our pain, but it also reflects an ongoing failure to address and resolve upsetting memories. Similarly, rather than providing a temporary escape, powerfully dissociative activities, such as playing video games, can become a way to avoid confronting problems in the real world. And, of course, the real world can't be avoided forever.

Is it possible to draw a line between healthy, temporary distraction from everyday stresses, and unhealthy, pathological dissociation? This is becoming an increasingly urgent question as technological advances make virtual worlds ever more convincing and immersive. Some players get so wrapped up in the alternative realities of video games they neglect their offline lives. Unable to regulate how much time they devote to play, their personal and work relationships suffer. They're addicted. In 2018, the World Health Organization (WHO) created a new mental health diagnosis of 'internet gaming disorder' in its revised International Classification of Diseases, though the move remains deeply controversial. There is a substantial body of evidence to suggest that the condition is associated with brain abnormalities similar to substance abuse disorders and pathological gambling.[5] But among psychologists who study online gaming there is a lack of consensus about how prevalent the problem really is, with some claiming that issues such as poor social skills, loneliness and low self-esteem *precede* excessive gaming rather than being its consequence. They argue that there is insufficient rigorous evidence to pathologize what, for the vast majority of players, is an enjoyable pastime that improves their overall quality of life.[6]

I suspect the jury will be out for years to come, with calls for more research from proponents on both sides of the argument, but some of the most interesting research to date into the balance of gaming's positive and negative effects has come straight out of Azeroth. For the benefit of readers who have never set foot on this fictional continent, let me explain a few basics. Players create avatars with particular skills and allegiances that progress to higher levels in the game by completing quests – usually involving defeating monsters and winning loot. In the process they become stronger, faster and better equipped with weapons, spells and potions that will allow them to tackle yet more perilous quests. Several years ago, an unlikely band led by Jeffrey Snodgrass, an anthropologist from

Colorado State University, travelled incognito among the goblins, night elves, holy warriors and dwarves of Azeroth. Like many anthropologists working in the field, Snodgrass and his four fellow players immersed themselves in this culture, taking detailed notes about their experiences and interviewing many of the characters they encountered. They also conducted a more formal, quantitative online survey of *Warcraft* players.

In this way, as the five battled their way up to the highest levels in the game, they gathered evidence about how it was affecting the well-being of players back in the real, offline world of careers and personal relationships. Their web survey, which garnered responses from 253 dedicated players, deployed two measurement scales: one to determine the degree of dissociation/absorption during play (experiences such as losing track of time, becoming unaware of events going on around them and identifying strongly with their in-game character); the other measuring addiction (for example compulsive play that harms relationships or job performance, symptoms of withdrawal when unable to play, and spending increasingly long hours in the game in an attempt to regain the pleasurable buzz that once came easily). Participants were also asked to rate the extent to which they believed playing *Warcraft* was impacting their stress levels, happiness and life satisfaction.

Almost half of all respondents reported that the game increased their happiness, and 64 per cent said it helped them combat stress, temporarily evading real-life problems with bills, their boss or partner, for example. As Richard, a man in his late thirties working in sales for a tech company, put it: 'Believe it or not, when I'm out there killing mobs and it's the same stuff over and over and over, my mind will wander. I'm relaxing. The game for some people is a retreat. It's a place where I can run and hide from real life.' Others likened the chores built into the game to meditation. A player in his early twenties said: 'You know honestly I like the repetitive activities that don't require a lot of thought, but that need to be done. I actually find them to be a calming influence... you know, a Zen-like state.'

However, around half were also showing classic signs of addiction, at least some of the time. They felt depressed, moody and nervous when they weren't playing, symptoms which went away when they returned to the game. They were having trouble cutting down. Sarah, a graduate in her mid-twenties, admitted:

> I have stayed up till 4 in the morning playing it. There is a level of immersion where you can keep playing it. You keep saying, 'One more thing and I'll be done' and then you forget you have to do things like sleep and eat. But once I start playing it's hard to tell whether or not I'll have the willpower to stop, particularly if I'm feeling stressed out or something.

Other respondents described losing friends and impaired job performance as a result of the amount of time they were devoting to the game. *Warcraft* addicts joke that it should be renamed 'World of Warcrack'. There are urban legends of players' desiccated bodies being found slumped in front of their screens.

But here's the rub: there was a statistically significant, positive relationship between the degree of dissociation players experienced in the game and the extent to which they felt they were addicted, and yet the amount of dissociation was also positively associated with increases in happiness and life satisfaction. So while increasing levels of dissociation led to a proportionate boost to players' mental well-being, it also put them at greater risk of addiction. 'The idea is that if you lose yourself, you escape,' Snodgrass told the press when his team's results were made public. 'But it is important to note that the escape must be controlled and temporary to be positive, so it leads to rejuvenation rather than simple problem avoidance – which in the end only increases the experience of stress.'

People who are already struggling with loneliness and low self-esteem in the real world seem particularly susceptible to the charms of alternative, online existences in which they are part of a comradely band of like-minded individuals engaged in heroic exploits. Games like *Warcraft* may provide them with a sense of achievement and self-worth that has eluded them in real life. 'Certain vulnerable players, we find, need and use WoW too much,' Snodgrass and his colleagues write.[7] 'Too much flow, it seems, can be dangerously addictive for these individuals.' And, of course, having lots of time on your hands can only make matters worse. One player interviewed by the anthropologists recalled a disastrous quest in which everyone in his team was annihilated: 'The leader just wigged out and he was on voice chat saying: "I swear to God, I only want people here who are like losers! I don't want people who have jobs or school!"'

Clearly the risks for vulnerable individuals need to be recognized and carefully managed, but Snodgrass and other psychologists who study gaming behaviour argue that health organizations like the WHO, by pathologizing gaming, risk throwing the baby out with the bathwater. A recent review of dozens of scientific investigations into the effects of interactive online games on children concluded that connecting with others to achieve common goals instilled a sense of meaning and purpose. Gaming also improved emotional stability, providing a safety valve for 'letting off steam' when the going got tough at school or at home. The studies reviewed did find that schoolchildren who were heavy video game users were slightly more likely to suffer from insomnia, anxiety and generally poor mental health than those who played occasionally, but those who *never* played the games appeared to be more at risk of these problems than either moderate or heavy users.[8]

Snodgrass and his fellow anthropologists think they know why. They write of their *Warcraft* study:

> The altered 'absorbed' states of consciousness that many gamers reach, even the powerful 'dissociative' identification some gamers have with their avatar, provide relaxation as well as some of the most satisfying, meaningful, and wondrous experiences of their lives. Even WoW addiction, we believe, should be considered too much of a good thing rather than simply a bad thing.

The intense concentration needed to reach the highest levels in a role-play game is accompanied by a pleasurable sense of flow, though players of *Warcraft*, *EverQuest*, *Final Fantasy* and other MMORPGs may be surprised to learn that as they sit before their screens slaying orcs and storm troopers they are doing much the same to their brains as cross-legged yogis focusing on their breath. The claim by Jayne Gackenbach, a psychologist at MacEwan University in Edmonton, Canada, isn't as outlandish as it sounds. Both video games and meditation are known to improve performance on tasks that require sustained attention or vigilance.[9,10] Meditators improve their capacity for vigilance by focusing on a particular stimulus – such as their breath or a mantra – for extended periods of time, noticing whenever their attention has wandered and gently bringing it back to the chosen object of attention. For action-adventure gamers, the price of a momentary lapse of concentration is rather higher: virtual death.

'What players are doing is training to be constantly vigilant of their environment, because they never know what's going to jump out at them,' Gackenbach explained to me. Both gamers and meditators appear to carry these improved attention skills with them into their dreams. As mentioned in the previous chapter, meditators are more likely to be lucid dreamers, and research by Gackenbach suggests people who spend a lot of their waking lives playing video games experience more frequent lucid dreams. When she compared meditators (including those practising contemplative prayer) with gamers, while the meditators ranked higher in terms of dream lucidity, gamers turned out to be particularly good at controlling the narrative of their dreams.[11,12] Games and dreams are virtual environments that subjectively can feel very similar, said Gackenbach. 'It's not rocket science. Players have had practice controlling artificial environments for hours and hours, of course they will do it in their dreams too.'

What gaming and meditation have in common is an intense, one-pointed concentration that evokes a sense of flow. In games attention is focused exclusively on the screen action whereas meditators focus all their attention on a particular sensory stimulus, such as changes in the sensations of breathing. But as we will see in Chapter 8, the underlying brain mechanisms may be very similar. Players enjoying this state of consciousness can become so absorbed that they lose track of time and fail to notice what's going on around them, often identifying more strongly with their avatar than with their everyday selves. This power of video games to create flow and loosen the bonds of selfhood has helped to reveal how the brain generates our feeling of embodiment, location and agency.

When researchers at Aachen University in Germany used functional magnetic resonance imaging (fMRI) to scan the brains of young men as they played a first-person shooter called *Tactical Ops: Assault on Terror*, the gamers' neural reward circuits lit up like neon signs as they picked off virtual terrorists.[13] The sensory and motor regions of their brains were also busy simulating the on-screen action – in the absence of any actual muscular contraction, of course, in much the same way these areas simulate the action in our dreams while our muscles are paralyzed. More importantly in the context of flow, activity in a part of their brains responsible for the feelings of physical embodiment was suppressed. So, as the gamers rampaged through the virtual world of *Tactical Ops*, looking through the eyes of their avatar, it was as if they had transferred a crucial part of their embodied selfhood into that world, leaving their real bodies behind. As they played, their brains were somehow contriving to ignore the sensory evidence that, in this instance, they were in actual fact lying flat on their backs in an MRI scanner.

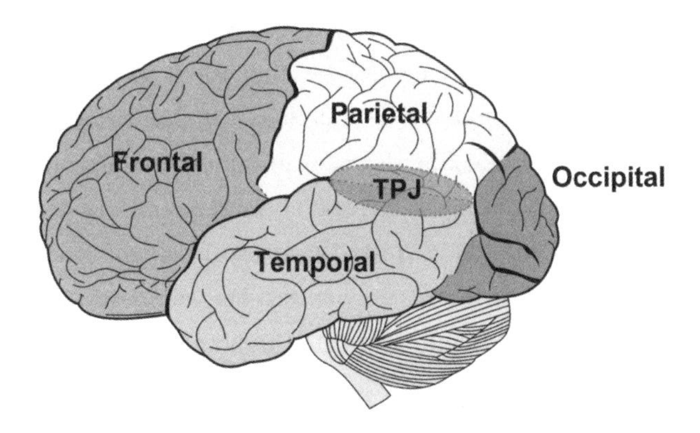


Figure 4: *The TPJ (temporoparietal junction) 'anchors' the self to the body by integrating sensory information*

The brain region identified in this experiment, part of the temporoparietal junction or TPJ close to where the temporal and parietal lobes meet (see Figure 4), integrates sensory information from our eyes, ears, muscles and 'somatosensory' touch sensors scattered across the body's surface.

This process of multisensory integration creates embodied selfhood: the conviction that, 'This is me, right here, an agent acting from inside this body.' Interestingly, the TPJ is also intimately involved in simulating what it's like to be somebody else – a cognitive skill known as 'theory of mind'. So it seems we use the same neural machinery to create our embodied self that we use to imagine what it would be like to be standing in somebody else's shoes. As a close neighbour of the somatosensory, auditory and visual cortices within each hemisphere of the brain – responsible for processing touch, hearing and vision, respectively – the TPJ is ideally placed to perform these complementary roles involving multisensory integration: simulating the self and the other.

Several independent lines of evidence have confirmed that the TPJ anchors the self to the body. When its activity is severely disrupted – not merely suppressed – the result is an out-of-body experience: the profoundly disorienting feeling of being located outside your body looking down on it from above. People who feel disembodied during a near-death experience will often interpret it later as evidence that they have a soul capable of existing independently of their bodies. Close brushes with death aren't by any means a prerequisite for an out-of-body experience, however. People who have epileptic seizures focused on the TPJ or brain damage in this area often experience them too, and they occasionally occur spontaneously in healthy people.

Feelings of disembodiment can also be induced and one way is to zap the TPJ of an awake patient during brain surgery. In 2016, the journal *World Neurosurgery* reported the case of a woman undergoing surgery at a hospital in Rotterdam in the Netherlands to remove a tumour. As surgeons 'mapped' the surrounding tissue using an electrode, which is standard procedure to avoid damaging tissue vital for sensation, movement and cognitive function, they electrically stimulated her left TPJ several times.[14] They write that the woman initially felt as though her right leg was being 'drawn toward the opposite wall of the operating theatre'. Then she had three full-blown out-of-body experiences in succession in which she thought she was floating just below the ceiling looking down at her body where it lay on the operating table.

It's worth emphasizing that despite decades of investigating disembodiment, thus far scientists have found no proof that a person can – without technological assistance – observe the world from a perspective other than that provided by their physical body.[15] A study published in 2014, for example, involved leaving images of religious symbols, animals and newspaper headlines on high shelves in hospital operating theatres and emergency departments that could only be seen from above. But while 9 out of 101 cardiac arrest survivors who were later interviewed reported having a near-death experience, none identified any of the images.[16]

Rather than proving the existence of disembodied minds, spirits or souls, out-of-body experiences add to evidence of the boundless creativity of the brain. Alongside phenomena such as the placebo effect, hallucinations and sensory illusions, these curious experiences are testament to our brains' ability to simulate every facet of consciousness – from the five senses to the feeling of being a self tethered to a body – not only in our dreams but also when we're wide awake. Only when the brain slips up does it become apparent what a fantastic job our biological virtual-reality simulator normally does.

If you're finding all this talk of disembodied selves a little disorienting, particularly if – like most people – you have never had an out-of-body experience, bear in mind that the sensation of occupying a body isn't a given of consciousness. Like pretty much everything else it has to be learned early in life and incorporated into the brain's working model of the physical world.

'The brain is of itself the model,' Karl Friston told me when I asked him to explain this counterintuitive idea. 'So all the connectivity, all the structuring, all the delicate wiring, constitutes a model of associations, a connectivity in there that's meant to emulate in some way what's out there.' The most important aspect of what's 'out there', said Friston, is our own body and how it operates and feels in relation to everything else in our environment. 'So there's a really challenging game on leaving the womb where you basically have to build a model of how your body responds, and what you can do and can't do.'

As a baby, you begin to construct this model of embodied selfhood by 'motor babbling', said Friston. You wonder, 'What would happen if I did that?' You randomly jerk your limbs, kick and clutch, just to see and feel the results, in exactly the same way you start learning how to speak by verbal babbling. By experimenting you begin to incorporate associations between your actions and inputs from the senses into your brain's virtual models. The process of embodiment and sensory entrainment had already begun in your final weeks in the womb as you jerked your limbs and as your eyes moved about beneath your eyelids, but after birth you were able to incorporate other objects and people into these basic models, for example by grasping your mother's finger or a rattle. 'This enables you to build concepts, such as there's "me" and "not me",' said Friston. 'There's "me" and "Mum". There's "me" and "an object: a rattle". You learn that this object has sensory consequences in both visual and auditory modalities.' The result of all this experimentation is an increasingly faithful, predictive model of your body and how it interacts with the wider world.

For the rest of your life, through multisensory integration centred on the TPJ, the model continues to anchor you to your body. Except, of course, during an out-of-body experience when a disembodied 'you' temporarily breaks free. You needn't be having a stroke, brain surgery or a near-death experience for this to occur, however. It is surprisingly easy to fool the brain into slipping its bodily moorings. In the classic 'rubber hand illusion', a subject sits at a table with her left forearm resting on the tabletop, hidden from her view behind a vertical partition. On the nearside of the partition, in full view, lies a lifelike rubber left hand. The experimenter then uses two paintbrushes to stroke the real and the fake hand in synchrony. This tricks the subject's brain into integrating the visual evidence of a 'hand' with tactile sensory inputs to her real, hidden hand to create the distinct impression that the rubber hand is part of her own body – so much so that if the experimenter without warning whacks the phoney hand with a mallet she will recoil in alarm.

The same kind of multisensory *mis*integration can be used to create a whole-body illusion. In order to identify the brain regions involved, the trick was recently performed on a man lying with his head inside an MRI scanner. A video displayed to the subject showed a lifelike male manikin seated facing away from him, its back being rubbed up and down with a plastic sphere the size of an olive on the end of a long stick (the person operating the stick was out of shot). In perfect synchrony with the stick in the video, a robotic arm built into the bed of the scanner simultaneously rubbed an identical sphere up and down on the subject's back. This delightfully eccentric mechanical set-up – which would make Wallace and Gromit proud – provoked an out-of-body experience in which the subject not only identified with the manikin in the video but also felt himself to be sitting where it was sitting. And the part of his brain responsible for the illusion was found to be the TPJ: its activity plummeted as it relinquished control over the integration of sensory information.[17]

Of course, we have been exploiting the phenomenon of multisensory misintegration to uncouple our selves from our bodies for almost three decades now, through immersive virtual reality. Even the relatively crude early technology for producing a sense of 'telepresence' was capable of triggering convincing out-of-body experiences. In 1990, at a robotics lab in Japan, the technology writer Howard Rheingold was among the first people to have the experience of donning a virtual-reality headset that tracked head movements and displayed video captured at another location. He describes turning and seeing a man who looked uncannily like himself. 'He looked like me and abstractly I could understand that he was me, but I know who me is and me is here. He, on the other hand, was there.' Looking at his own body through the eyes of a robot, he no longer felt any sense of occupying or owning it.[18]

Today, anyone with enough money can get a taste of totally immersive virtual reality through commercial gadgets such as Microsoft's Oculus Rift headset, which transports users into a virtual world through audio and video inputs synchronized with their head movements and changing location within a room. A controller for each hand also allows them to interact with the virtual environment in distinctively human ways, such as shooting a gun and grasping. The technology may be relatively new, but in the brain the TPJ performs exactly the same job as it always has, conjuring selfhood from synchronous multisensory inputs.

Apart from taste and smell, perhaps all that's lacking from this whole-body illusion are 'haptic' touch simulators to deliver virtual blows and caresses – though these are sure to follow. In principle, twenty-second-century technology could substitute every sensory input to provide the ultimate in dissociation: total liberation from the limitations of the human body. Imagine for a moment a distant future in which a disabled interplanetary biologist can transfer his bodily presence into a genetically engineered cyborg that can fight, fly and even fall in love with an alien – as portrayed in the 2009 film *Avatar*. Assuming there are skilled technicians on hand to maintain the virtual-reality gear and cater for the needs of the human's biological body, the only hitch might be motion sickness, caused by dissonance between the cyborg's visual inputs and what the man's vestibular balance system is telling his TPJ about his head's *actual* position and movements.

Back down to earth in the twenty-first century, virtual-reality technology has been providing promising new opportunities for psychotherapy, for example allowing people with anxiety disorders and phobias, such as an excessive fear of heights, to confront and conquer their demons within safe, controlled virtual environments. Virtual-reality-assisted therapy has even scored some successes in psychosis, helping patients overcome their suspicion of strangers.[19–21] That patients are perfectly aware intellectually that what they are experiencing is an illusion and their safety is not actually threatened doesn't seem to matter. In virtual reality, as in everyday reality, the processes of learning and unlearning conditioned responses to fear-provoking stimuli is automatic and unconscious.

In theory, games could take this to a whole different level. The ultimate goal for psychiatrists will be to create immersive games for mental illnesses, such as depression, anxiety and schizophrenia, that are not only therapeutic but also entertaining, maximizing benefits by motivating patients to put in the necessary hours playing them. Unfortunately the Internet-based therapeutic games developed thus far, oxymoronically known as 'serious games', have high attrition rates. In studies of games designed to combat depression and reduce binge drinking, for example, the vast majority of adolescents who played failed to complete the programme.[22]

Perhaps the therapeutic games our grandchildren will play will draw their inspiration from addictive MMORPGs such as *World of Warcraft* and *Final Fantasy*, which keep players interested not only with epic storylines and high production values, but also through precisely calibrated 'reward schedules' that allow rapid progress through lower skill levels, only becoming seriously challenging at higher levels. These mind-healing games will give players the ability to communicate with each other in real time, building a supportive social network that will help reduce the stigma attached to mental illness through sheer force of numbers. Of course, as with any social network, provision will have to be made to police antisocial behaviour if it arises.

The best therapeutic games will also incorporate biofeedback, responding to changes in physiological variables, such as breathing and heart rates, skin conductance and temperature, that reflect players' emotional state. Pilot studies of a games platform called PlayMancer developed with funding from the European Union, have shown promising results. Used in conjunction with standard cognitive behavioural therapy (CBT), it is designed to improve the ability of people with conditions such as gambling disorder and bulimia nervosa to control their impulses. In a game called *Treasures of the Sea*, for example, players collect precious artefacts from the seabed while harvesting the oxygen they need to stay underwater from passing 'balloon fish'. If the equipment detects that they are getting stressed or frustrated, the fish become more difficult to catch and the player is taken to a relaxation zone where they learn how to calm down. Conversely, the game rewards players whenever they manage to keep their cool.[23]

Biofeedback games platforms like PlayMancer require specialist laboratory equipment, but others in development can be played at home using off-the-shelf devices such as the MindWave headset, which monitors brain activity using 'dry electrodes'. In the lab, a sticky gel is needed to ensure a good electrical contact between electroencephalography (EEG) electrodes and the scalp, but the makers of MindWave say their device gets round this by using signal-boosting algorithms to cancel electromagnetic noise from nearby electrical equipment. By detecting alpha waves in the brain, which correspond to relaxed states of mind, the headset provides biofeedback for games designed to help patients control anxieties about things like driving, flying or social situations. The games – which have yet to undergo clinical trials – combine classical exposure therapy for phobias with challenges aimed at developing emotional regulation strategies. In an immersive children's game called *Mindlight*, for example, a character called Little Arthur tries to rescue his grandmother from a dark, haunted mansion. He finds and puts on a magical headset with a lamp that shines progressively brighter as the player learns to relax, revealing shadowy phantoms to be harmless objects like wardrobes or grandfather clocks.

If research findings by scientists such as Snodgrass and Gackenbach are confirmed, successful commercial games already deliver many therapeutic benefits. In addition to reducing stress, increasing emotional stability and fostering online friendships, sense of meaning and accomplishment, by boosting dream lucidity and control they may also help players with post-traumatic stress disorder (PTSD) combat the recurrent nightmares associated with past traumatic events.

When Gackenbach analyzed the dreams of US military personnel who had served their country in war zones such as Iraq and Afghanistan, unsurprisingly they often had violent themes.

But some telling differences emerged when she compared regular gamers with occasional gamers.[24] 'The heavy gamers were fighting back,' Gackenbach told me. They were more in control of what was happening in their dreams. 'Yes, there were bombs going off, there was blood everywhere, but their gun worked and they could respond, whereas the low gamers' trigger pull was too heavy, they felt frozen, they couldn't move.' In other words, regular gamers were less prone to the utter helplessness that turns merely bad dreams into terrifying nightmares.

Underlying the power of virtual-reality games to channel our behaviour and emotions in more healthy directions is the convincing illusion of being physically *present* in a fantasy world, thanks to the synchronous, multisensory integration performed by our brains. Unlike reading a novel or watching a film, no matter how gripping these might be, virtual reality provides that crucial sense of personal involvement. Even though we know, as in a lucid dream, that our adventures are pure fantasy, the brain learns nonetheless, adapting its predictive models of the world as though the experiences were real. As a result, gaming is not only therapeutic but also provides a liberating sense of empowerment. 'If you work as a delivery guy you don't get to fight off villains very often!' joked Gackenbach. 'People live to game. With work it's all about bringing home the bacon, whereas in a game they have real power, they feel like they're having adventures and they're part of a community.'

Feeling part of a community may be as important as the thrill of battle. Online 'gaming clans', as they are sometimes known, provide a shared sense of belonging, meaning and purpose. In an individualistic society, Gackenbach and Snodgrass believe fantasy role-play games like *World of Warcraft*, with their Jungian archetypes of bands of heroes embarking on quests to defeat monstrous enemies, serve much the same function in modern, technological cultures that mythological stories once did in tribal societies.[25,26] They are our bridge to an idealized, highly absorbed, collective state of consciousness that transcends the limitations of narrow self-hood – not unlike trance states, which are the subject of the next chapter.

Perhaps it is no coincidence that the hit TV series *Game of Thrones*, the hugely popular multiplayer game *World of Warcraft* and its forerunner *Dungeons & Dragons* all drew their inspiration from the mythical landscape of 'Middle Earth' created by J. R. R. Tolkien, who in turn drew his inspiration from Nordic and Germanic folklore. When my schoolmates and I played *Dungeons & Dragons* we considered it a guilty pleasure, but the self-transcendent appeal of these archetypal fantasies may be universal.

# 4

# *Puppets on a String*

On a muggy summer's evening in 1969, people across Britain witnessed something astonishing. Through the windows of their TV sets they were drawn into a world in which the spirits of gods, animals and inanimate objects can take over the bodies of human beings. The scenes were shot after dark on the other side of the world, in remote mountain villages on the Indonesian island of Bali. Describing the unsettling spectacle were the reassuring tones of a young David Attenborough. He explained that these were ancient, animist rites that had survived the arrival of Hinduism from Java 400 years previously.[1]

In the first scene, two girls aged eight or nine years kneel on either side of a shaman dressed in white robes and a turban. The girls are clothed in lavish, brightly coloured costumes with elaborate crowns on their heads decked with white and yellow flowers. Their faces have been whitened to give them a ghostly, doll-like appearance, but otherwise they are perfectly ordinary children. Occasionally they stare distractedly into the camera lens, meeting viewers' eyes. In the shaman's hands are two wooden puppets which he suffuses with fragrant smoke from smouldering leaves, just as he would with human dancers to send them into a trance.

The puppets are then hung from a line strung between two poles held upright on either side by the girls.

Now the puppets are brought to life, jerking up and down as the girls pound the poles rhythmically on the ground. Sitting behind them, female villagers sing, entreating the gods to come down and animate the girls, just as the girls are animating the puppets. 'Slowly the children's eyes fall shut,' Attenborough intones. 'Slowly they drift into trance.'

The villagers now address the girls as *Ratu*, Your Highness, because *Sanghyang* or 'god spirits' have taken control of their bodies. 'The children are acquiring new characters,' says Attenborough. 'They may become petulant and arrogant, and quite unlike their normal selves.' The shaman makes an offering, then the girls begin the *Sanghyang Deling* or 'divine dance' to sanctify the village. Climbing onto the shoulders of two men, their eyes still firmly shut, they stand up and begin to dance unsupported, effortlessly graceful and poised as the men walk in a circle. Perfectly balanced on the men's shoulders, the girls dance like expressionless puppets. Attenborough explains:

> In times of famine or pestilence when disease threatens, the children may be called upon to perform every night for weeks. They must be children, for the heavenly spirits wish to inhabit pure bodies uncorrupted by adolescence. And they must dance like puppets, impassively and in unity, for they are portraying the puppets of the gods.

Some time later, their dance nearly over, the girls jump down to the ground. But the gods will not release them yet. First they must dance barefoot through the embers of a fire made from coconut husks, which they do impassively and without flinching or opening their eyes even once. In trance, we are told, the Balinese can perform acts they couldn't possibly repeat without pain or indeed at all in a normal state of consciousness.

When the time finally comes to restore them to their former selves, they sit blank-faced on either side of the shaman. He prays, then uses a flower to sprinkle them with holy water. Still absorbed, in a state of deep relaxation, the girls' heads sway slightly as drops trickle down their whitened cheeks. At last one girl and then the other opens her eyes. 'The gods are leaving them, the trance is over,' says Attenborough. 'Once more they are little girls. Once more they have demonstrated that the gods are all around, invisible but in direct communication with the people.'

In a later scene, viewers witness rituals in neighbouring villages in which men are possessed by the spirit of pigs. On all fours, they charge about wildly as onlookers jeer and jostle them. To the great amusement of the villagers, the shaman sends one man into a trance in which his body is taken over by the lid of a pot. This man runs here and there with the lid in one hand, manically slamming it onto any flat surface he can find. Unlike the *Sanghyang Deling*, this is entertainment, but the trance state is nonetheless perfectly real and can be dangerous. Trance endows superhuman strength, and when it's time for the men to be released it takes six or seven others to restrain them.

The possessed are not play-acting. One night, Attenborough says, a man in a pig trance broke through the cordon of villagers and escaped into the surrounding darkness. He was found the next morning in a mud wallow and was sick for days afterwards, 'for in his pig incarnation he had eaten dung, as pigs will'.

What Britons saw that night in 1969 – perhaps most poignantly the dancing girls possessed by gods – gave an unnerving insight into the true nature of human consciousness. A month later, many of the same viewers will have watched spellbound as another pair of faraway, ghostly figures danced weightlessly across the surface of the moon. The scenes they witnessed in Bali were equally extraordinary. Half a century on, psychologists are still coming to terms with the implications of trance. Many readers will be familiar with trance as entertainment, in which a stage hypnotist persuades audience members that they have lost the ability to read or talk, that colour pictures are black and white, that they are competing in a Mr Universe contest or possessed by the spirit of Elvis. Charismatic priests deploy the same techniques for the edification of their congregations when they convince believers that their rheumatism has been cured, their tinnitus has vanished, that they can throw away their spectacles or walking sticks: God has put forth his power and healed them.

The permutations are as varied as human experience itself, because what these apparently miraculous transformations prove is not the existence of gods or the spirit realm, but that where consciousness is concerned, for much of the time, we get what we expect. As we've seen, neuroscientists now believe consciousness to be a mostly top-down performance in which preconceptions play the leading roles. In hypnosis and trance, these preconceptions are shaped by the power of suggestion. To stretch my company HQ analogy a little further, in these altered states it's as if the highly paid executives who have their offices on the top floor have gone to lunch, leaving charismatic but capricious management consultants in charge. 'Just do whatever they tell you,' were the bosses' departing words. For the limited time they are away, the consultants' crazy ideas hold sway as staff willingly suspend their critical faculties and natural scepticism.

Among the suggestions to which people can succumb are experiences of anaesthesia, paralysis, amnesia, false memories, involuntary movements and hallucinations. A common feature is a sense of losing control over your actions and sensations, perhaps being unable to lift your arm, seeing words as colours or experiencing an unexplained taste in your mouth. In the popular imagination, to cause outlandish changes like these would require supernatural powers or great skill and will probably involve sending the person into a deep, semi-conscious trance state, whereas in actual fact psychologists studying hypnosis routinely evoke these effects in subjects who remain fully aware in an apparently ordinary state of consciousness. Faith, expectation and atmospherics undoubtedly play powerful roles, but it's surprisingly easy to seize executive control.

In the brain, the 'top floor' from which this control is normally exercised is the frontmost portion of each hemisphere's outer surface or cortex, known as the prefrontal cortex. A subdivision of this on either side of the forehead, the dorsolateral prefrontal cortex, or DLPFC, is particularly important for executive control. As you may recall from Chapter 2, the DLPFC is deactivated in dreaming sleep but stutters back to life in lucid dreams when the sleeper realizes that what he or she is experiencing is pure fantasy. This part of the prefrontal cortex is essential for metacognition: for performing tasks in which the mind must become aware of its own operations. Several studies have demonstrated that, in their everyday state of consciousness, people who are highly prone to hypnotic suggestion perform less well on tests that make calls on the DLPFC, such as activities involving working memory and selective attention. Interestingly, damage to this region as a result of a stroke is associated with surreal beliefs such as the Capgras delusion (the belief that your significant other has been replaced by an imposter) and the Fregoli delusion (the belief that someone you know is following you around in a variety of disguises).

It may come as no surprise then that if scientists disrupt volunteers' DLPFC in the lab, this makes them more prone to suggestion. In one experiment, when researchers at the Sackler Centre for Consciousness Science, part of the University of Sussex in the UK, pulsed this region of subjects' prefrontal cortex with a weak magnetic field – known as 'repetitive transcranial magnetic stimulation' – it made them more susceptible to a range of suggestions. For example, it became easier to convince them that their arm could levitate of its own accord; that it had become so stiff they could no longer bend it at the elbow; or that there was suddenly an unexplained sour taste in their mouth. When the researchers applied magnetic pulses to another part of the brain called the vertex, which plays no part in metacognition, this didn't have any effect on participants' suggestibility.[2]

Priests and hypnotists obviously don't use magnetic fields to disable the prefrontal cortex of believers and audience members, but they do deploy a purely metaphorical form of magnetism. Charisma and people's prior beliefs about their powers are key: if you're told that someone is renowned the world over for their abilities as a hypnotist or spiritual healer, your brain will be more likely to surrender executive control to them.

In 2011, anthropologists at Aarhus University in Denmark conducted an experiment in which they scanned the brains of two groups of people – Christians who believed in the healing power of prayer and atheists who did not – as they listened to prayers of intercession, asking God to heal the sick. The subjects were told that the prayers would be spoken by one of three types of person: a non-Christian, a Christian or a Christian known for his healing prowess. In actual fact, all the prayers were read by 'ordinary' Christians. When believers thought they were listening to a non-Christian, there was a significant increase in DLPFC (executive) activity, but when they thought they were listening to a Christian famous for his healing abilities there was a massive decrease in activity. When they thought they were listening to a run-of-the-mill Christian, levels of activity fell somewhere in between these extremes. Among unbelievers, however, there were no significant differences between the three conditions. Interestingly, when believers were asked to rate each speaker after the scans, the greater the reduction in executive activity that was observed in their brains, the more highly they rated the speaker's charisma.[3]

There remains some disagreement over whether it's the DLPFC in the right or left hemisphere of the brain that exerts the kind of executive control that shamans, priests and hypnotists hijack – experiments that disrupt activity in either hemisphere using magnetic fields have yielded conflicting results – but psychologists are in broad agreement about the overall mechanism. They divide the process into two stages (though it can sometimes be difficult to differentiate them). The first, 'induction' stage often involves focusing the subject's attention, relaxing and reassuring them that they are in safe hands. In a shamanic or religious setting, the ritual will be highly absorbing and led by a charismatic, authoritative individual who commands respect. The context in which the procedure takes place, the words used and the subject's beliefs will also play a pivotal role in priming them for what they are about to experience. Even saying the word 'hypnotism' during a hypnotic induction, for example, has been found to increase a person's suggestibility. In the second stage, the suggestion itself will be delivered. This will either be stated explicitly or will be implicit, perhaps as a result of participants being in a religious setting following a well-known, shared cultural script.

Under hypnosis or in a trance, if someone asks you to perform a particular action or imagine a particular sensation, it will seem as though these things are happening of their own accord – as if you were merely a puppet and someone else was pulling your strings. There will be a lack of self-awareness or metacognition, a disconnect or dissociation between willing something to happen and being aware that it is you who is willing it. The effect can be enhanced by phrasing suggestions to imply this lack of agency. So for example rather than saying 'You will raise your hand,' a hypnotist will say 'Your hand will rise.'[4]

When highly hypnotizable individuals are given a suggestion that an external force is controlling their movements, for example – much like a Balinese ritual involving 'possession' by a god or the spirit of an animal or inanimate object – activity in the part of their brain responsible for the sense of motor agency or control will be inhibited.[5] Similarly, a suggestion of disembodiment – an out-of-body experience – will be accompanied by reduced activation of the temporoparietal junction (TPJ), which as we saw in the previous chapter helps create the feeling of inhabiting a physical body through a process of multisensory integration.[6] Crucially, the brain activity of highly suggestible people as they respond to a hypnotic suggestion looks no different than if they were experiencing the real thing: they're not simply playing along. So if a hypnotist tells them they can see colour in a black-and-white image, or only shades of grey in a colour image, this boosts or suppresses activity, respectively, in the colour-processing regions of their cortex, as if the pictures really had changed.[7]

Piecing all this together, the inescapable conclusion is that people in a hypnotic trance or listening to a charismatic healer are not obligingly playing a role. For them, it's really happening. This is more evidence that conscious experiences are powerfully influenced by top-down expectations, including those parachuted into our brains by skilled, renowned or charismatic individuals. You could be forgiven for thinking that the take-home message here is to steer well clear of such folk. But to reject these uncanny effects outright on the grounds that they are 'not real' would be as misguided as refusing to watch a film on the grounds that the pictures on the screen 'aren't really moving'.

Whether we realize it or not, people in positions of authority who command our respect – such as our favourite politicians, or our parents and teachers when we're young – deploy the power of suggestion to change the way we experience the world. Similarly, good doctors use their charm and presentation skills to maximize the placebo effect, making their drugs and other treatments as efficacious as possible. If they can persuade us to believe in their medicine, it will be more likely to work. Faith really can heal. Humans have doubtless been harnessing suggestion for the purposes of physical and psychological healing for millennia. In a religious or shamanic context, particularly in communities beset by hunger, disease or poverty, or a feeling of powerlessness in the face of their enemies, rituals led by charismatic authority figures provided a respite from the hard realities of life and a renewed sense of hope and purpose.

Science only really got in on the act in the nineteenth century. In 1842, the Scottish surgeon James Braid coined the term 'neuro-hypnotism' ('nerve sleep') to describe the phenomenon. In so doing he was going against the received wisdom that the effects were mediated by an invisible, physical force or 'animal magnetism' between the therapist and patient, in favour of a purely psychological explanation. Later, when he discovered that sleep wasn't involved either, he perceptively renamed the process 'monoideism' ('single thought' or absorption), though perhaps unsurprisingly the term never caught on.

For much of the twentieth century hypnosis remained on the fringes of medicine, but in recent decades its reputation has revived as clinical studies provide support for its use to enhance the efficacy of conventional treatments. There's good evidence, for example, that hypnosis helps people lose weight when combined with cognitive behavioural therapy (CBT) for obesity. There are also multiple clinical trials suggesting it can help treat irritable bowel syndrome, chronic pain and depression, that it can alleviate nausea and vomiting in chemotherapy, and children's fear of needles. Most impressively of all, hypnosis has been used to replace anaesthesia during some forms of surgery.[8]

Hypnotherapy has become particularly popular as a way to break up ingrained habits and fears, for instance to quit smoking or overcome a phobia. A few years ago, I turned to a hypnotist when my fear of flying threatened to ground me completely. I'd always been an anxious flyer and, like many people, often resorted to a drink or two before a flight to settle my nerves. Matters came to a head, however, after an alarming experience on a flight to Peru. The plane on which I was a passenger aborted its landing at Jorge Chávez International Airport in Lima at the last possible moment – engines roaring as it fought to regain altitude – when the pilot spotted another aircraft blocking the runway. We landed safely on our second approach, but as luck would have it the next time I got on a plane a few months later, a budget flight from Amsterdam to Birmingham, my anxieties were ramped up another notch when, shortly before takeoff, one of the cabin crew asked if I would mind moving to a vacant seat next to an emergency exit. She casually explained that if the plane ditched in the sea or made an emergency landing everyone would be relying on me to operate the door.

I'd be the first to admit that, like any phobia, excessive fear of flying is irrational. The previous year, 2017, there hadn't been a single jet passenger aircraft fatality. Globally, around 1.25 million people annually die in road traffic accidents, roughly equivalent to a Boeing 747 falling out of the sky every hour or, as I tried telling myself, two of the dinky planes on which I was flying. This didn't help and when I got back home from Amsterdam, despite no unscheduled landings, I wondered whether I'd have the guts ever to fly again. Spurred into action by this depressing thought, I sought help.

Pamela, my hypnotherapist, didn't conform to any of the stereotypes. She didn't swing a fob watch inches from my face or ask me to look into her eyes. She didn't murmur 'You are feeling very sleepy... very sleepy...' Rather, after we had discussed my negative flying experiences, she instructed me to focus on my breath, relax, then imagine lightning flashing backwards and forwards and from side to side across my brain. I was told to intersperse these imagined lightning bursts with picturing myself on a plane strapping myself in ready for takeoff, then being transported to somewhere I had felt perfectly safe and contented. Pamela's voice was calm and authoritative as she rapidly talked me through this sequence several times in succession. I was sceptical. But after a single session she declared me cured, and so far it seems to have worked. I haven't had any qualms about boarding an aeroplane since.

The goal of any kind of psychotherapy is to repair faulty, maladaptive cognitive models of reality. My faulty model paired flying with catastrophe – an association that Pamela sought to uncouple and substitute with an association between flying and my 'happy place'. Incidentally, you can save yourself the expense of hypnotherapy and hypnotize yourself by following the guidelines at the end of this chapter (see pages 110–114). To understand why hypnosis and self-hypnosis work, it helps to know how the brain creates the conscious experiences that are emotions. When you suffer from a phobia, a particular sensory stimulus, such as the sight of a spider or the mere thought of one, will automatically flood your bloodstream with adrenaline, resulting in a pounding heart, rapid shallow breathing, sweating, tensed muscles and other signs of physiological arousal. But it's important to realize that adrenaline can provoke any of several emotions – not only fear but also elation, anger and excitement – which all have identical physiological signs. How can the same physiological changes evoke such different emotions?

The answer once again is *expectation* and it helps to explain how hypnosis works, because if you can change someone's (largely unconscious) expectations you can change what they experience. In a classic experiment in the early sixties, the American psychologists Stanley Schachter and Jerome Singer managed to evoke either anger or euphoria in male volunteers experiencing identical physiological changes, simply by manipulating their expectations. They first informed subjects they were going to inject them with a vitamin to test its effect on vision. In actual fact the shots contained either adrenaline or a placebo (saline solution). Some were told, accurately, that the injection might have temporary side effects such as a pounding heart and facial flushing; others were told, inaccurately, that it could cause itching or numbness; and a third group were not told anything about the effects.

After administering the shot, a researcher left each subject filling out questionnaires for twenty minutes with someone else who had also just been injected. In reality, this other person was a confederate of the researchers who was there to create an expectation either that the 'vitamin' created a feeling of euphoria or that it provoked anger. To do this the stooge either messed about joyously – making and flying paper aeroplanes and twirling a hula hoop he found in a corner of the room – or complained endlessly about the whole experiment, becoming increasingly irascible. Objective ratings by a hidden observer, and the subjects' own subsequent reports, revealed that for those injected with adrenaline and informed accurately about the physiological effects, the stooge had little influence on their emotional state. But for those who were misinformed about the effects or told nothing, they were strongly influenced by the stooge – if he acted euphoric they felt euphoric and if he acted angry they felt angry. In other words, in the absence of an adequate explanation for their physiological state, the subjects took their emotional cue from the other person.[9]

My hypnotherapist sought to change my expectations about flying so that when my heart began pounding the next time I got onto a plane, my brain would interpret this as holiday excitement rather than fear of flying. After all, the physiological signs are exactly the same. All that needed to change was my interpretation of those signs. Anil Seth, co-director of the Sackler Centre for Consciousness Science, has dubbed this 'active interoceptive inference'. I'll explain the 'active' part in a moment, but the overall thrust of the theory is that our emotions are the product of top-down, cognitive inferences about the causes of signals from sensors in our muscles, heart, lungs and other organs (collectively known as 'interoception'). So our expectations and the environmental context of physiological changes are what determine our mood, rather than the changes themselves.[10]

When I called Seth up to ask him to explain, he told me:

> A single initial physiological change such as a racing heart can result in different emotions depending on context. So if I'm in a context that is very comfortable and happy then for the same basic physiological change I'll have a high-level

> prediction that this is a sign of good things going on, whereas if I'm in a scary situation I will have a high-level prediction that this increasing arousal should be interpreted as bad things and I should run away – in other words an emotion like fear.

His theory fits neatly within the predictive processing framework of perception we encountered in the context of 'exteroceptive' (outer) perceptions like vision, touch and hearing. According to this view, the job of the brain is to infer the hidden causes of its sensory inputs. Predictions are passed downwards through its information-processing hierarchies and prediction errors are reported back in the opposite direction (see Figure 2, page 45). When predictions from the higher, cognitive levels fail to explain sensory data coming up from lower levels – when there are unresolved prediction errors – the brain has two options. It can either select a different model that is a better match for the sensory data, which is the basis of perception ('That must be a plane, not a bird as I first suspected') or it can allow predictions to pass up through the hierarchy, updating its models to explain the new data, which is the basis of learning ('It's a flying man in a cape'). In other words, the object exciting your senses is either something you've seen many times before, or something unprecedented that demands a completely new explanation.

But there's a third option. Refinements to predictive processing theory over the past decade have added a whole new dimension, encompassing not only perception and learning but also action. The idea is that when an organism encounters prediction errors it can also resolve them by changing its sensory inputs. So, for example, you might turn your head and hold your hand up to shield your eyes from the sun to get a better look at this unidentified flying object. In 2010, Karl Friston and his colleagues at University College London proposed that this is in fact how our muscles are controlled.[11] They call it 'active inference'. Their radical proposal is that the central nervous system doesn't issue motor commands to perform some action, as neuroscientists once thought. Rather, it predicts the sensory signals it will receive from stretch receptors in muscles and tendons as a result of the desired action. These predictions are then fulfilled automatically by spinal reflexes, which move our muscles in such a way that they minimize the ensuing prediction errors. According to this view, you move your arm not by commanding it to move but by continually predicting how it will feel as it travels to where you want it to be. In a very real sense, your arm's motion is a self-fulfilling prophecy.

Even if you're doing something as deceptively simple as walking, let alone playing tennis or driving a car, it's easy to see how this would be more efficient than trying to exert conscious, real-time control over the activity of each and every muscle by issuing a stream of precise motor commands. To extend my metaphor of the hierarchy of a multinational company, rather than the CEO giving a blizzard of highly detailed orders to her subordinates, she simply outlines her expectations about where the company should be – her vision for the future – and says 'Make it so.' Branch managers at retail outlets across the world then scramble to turn her vision into reality using their local influence and expertise.

Anil Seth believes that an active, inferential mechanism also drives homeostasis, continually adjusting variables such as temperature, heart rate and blood pressure to maintain optimal internal conditions. In this case, predictions are enacted by the autonomic nervous system (the involuntary, automatic division of the peripheral nervous system). According to Seth, part of the brain called the anterior insula, hidden deep in a fold of cortex between the brain's temporal, frontal and parietal lobes, issues predictions about the body's physiological state. Like the setting on a thermostat, these are the optimal 'set points' appropriate for particular circumstances, as determined by the brain's models of the internal world of the body. The predictions are then fulfilled by autonomic reflexes.

'Predictions can be used for control,' says Seth. 'And control is not just an added extra; it's fundamentally why we have predictive models of perception in the first place.' He sees perception not as the passive harvesting of information from our internal and external environments, but intimately bound up with all the activities that keep us alive. 'When it comes to the insides of our bodies, heart rate, blood pressure, all these important physiological variables, what matters is not figuring out what they are or where my internal organs are but *regulation* – keeping these variables within the tight bounds that are compatible with survival.' If he is right, emotions are the conscious component of these predictions that take into account our expectations and the environmental context of whatever physiological changes have been detected. Strong support for the theory comes from studies showing that the anterior insula integrates interoceptive (internal) and exteroceptive (external) sensory inputs and is involved in every kind of emotion, from anger, fear and hatred to lust and jealousy.

In recent years the mind has been shown to play a much more active role in regulating our physiology than biologists once thought was possible. We've seen that rats deprived of the neural 'reboot' that sleep normally provides start to lose control over physiological variables, such as body temperature, eventually resulting in their death.[12] A more subtle but equally unexpected influence of the mind over body temperature becomes apparent when our sense of body ownership is disrupted. During the rubber hand illusion described in the previous chapter (page 79), in which people experience an artificial hand as part of their own body, the temperature of their real hand plummets as interoceptive attention is withdrawn from it.[13]

Some highly unusual individuals have even learned to control their body temperature at will. You may recall that the endurance swimmer Lewis Pugh can raise his core temperature before plunging into freezing water (see page 33). Tibetan monks do something similar, using a meditative technique they call *g Tum-mo* to warm themselves. According to an anecdotal report by the travel writer Alexandra David-Néel in the thirties, *g Tum-mo* was practically a competitive sport among novice monks:

> The neophytes sit on the ground, cross-legged and naked. Sheets are dipped in icy water, each man wraps himself in one of them and must dry it on his body. As soon as the sheet has become dry, it is again dipped in the water and placed on the novice's body to be dried as before. The operation goes on in that way until daybreak. Then he who has dried the largest number of sheets is declared the winner of the competition.[14]

It goes without saying that anecdotal reports like this should be treated with scepticism. Nonetheless, there is scientific evidence of a less dramatic effect. Writing in the journal *Nature* in 1982, researchers from Harvard Medical School described a study of three middle-aged monks who lived in unheated, uninsulated stone huts in the foothills of the Himalayas near Dharamsala in India, at altitudes of between 1800 and 2800 metres. In the study, which was only made possible because the Dalai Lama himself persuaded the monks to take part, the scientists monitored changes in skin temperature before, during and after the meditation. They recorded increases of up to 8.3°C in the monks' fingers and toes as they meditated that were sustained for around thirty minutes – easily exceeding temperature increases previously recorded by other scientists in the fingers of people as a result of hypnotic suggestion.[15]

That the monks and the athlete have somehow learned how to turn up the setting on their bodily thermostat is remarkable, though perhaps 'superhuman' feats like these shouldn't surprise us any more; if the theory of active inference is true, top-down predictions are fundamental to how we control our bodies. They also allow shamans, hypnotists, doctors and priests to get in on the act, because by manipulating our predictions (our expectations) they can change our perceptions, our physiology and our behaviour. We label these spooky effects hypnotic trance, the placebo effect, faith healing, positive psychology and so on, according to the circumstances, but the underlying mechanism is rooted in the biology of active inference – the way creatures with sophisticated nervous systems change themselves and their environment in order to survive and reproduce.

Crucially, the various dimensions of our feeling of selfhood may also arise through active inference. For example, according to Anil Seth, when the autonomic nervous system successfully fulfils the brain's predictions about a physiological change, this gives rise to a feeling of embodied selfhood or 'presence'. Similarly, when spinal nerves successfully fulfil predictions about, say, turning our head, it gives rise to a feeling of personal agency. If for whatever reason there is a failure of communication between my brain and my body, however, this will cause confusion about whether *I* am the originator of my actions or whether my thoughts, perceptions and actions are being controlled by some external agent – a god or another person with special powers, perhaps. In some contexts we might be tempted to attribute this to schizophrenia, in others we'd put it down to hypnotic trance.

Of course, the actual powers of healers and hypnotists are strictly limited. Faith in a healer or a medicine, no matter how robust, can't fix a severed spinal cord or cure a terminal illness. It's also worth bearing in mind that the degree of hypnotic suggestibility varies greatly from person to person. And even among those who are highly suggestible, trance states eventually wear off as reality reasserts itself: a healthy brain can only maintain inaccurate hypotheses about the world for so long before they collapse in the face of contradictory evidence – the escalating prediction errors being reported back by the senses. In the aftermath of a stage show or an animist ritual, sooner or later you will realize that you are no longer possessed by the spirit of Elvis, a pig or the lid of a pot. Nonetheless, that such faulty models of reality can be maintained in the teeth of contradictory evidence for as long as they are surely demands an explanation.

As we've seen, the 'secret sauce' that all these rituals and practices seem to have in common is absorption: an exceptionally narrow focus of attention that temporarily excludes everything else from consciousness. According to prediction error processing, attention is the gatekeeper determining whether or not a prediction error is allowed to pass upwards through the processing hierarchy and change our cognitive models. This is because the brain is not only in the game of predicting the hidden causes of its sensory data, but must also predict the reliability or 'precision' of that data. Attention and action (think of the way your eyes dart here and there as you take in a visual scene) are driven in tandem by this restless search for reliable information.

Psychologists studying hypnosis have theorized that because the brain has a finite capacity for conscious attention, streams of sensory evidence that would normally contradict the hypnotic suggestion get overlooked as a result of absorption. Viewed through the lens of predictive processing, sensory prediction errors fail to penetrate the outer gates of the brain's processing hierarchy because the 'gain' on these signals has been suppressed while our attention is elsewhere. Put another way, when the volume of a particular channel of prediction error signals is turned up to the max, the likelihood of signals getting through on other channels is greatly reduced.

This may be what happens during the absorption of trance and in deep meditative states, when particular sensory inputs are temporarily suspended. It results in dissociation – a disconnection from the reality checks normally afforded by ordinary, less highly focused attention to our outer and inner worlds. A healthy degree of dissociation from reality is part and parcel of all kinds of entertainment, not least video games, and there's more on dissociation and meditation in Chapter 8. But the phenomenon can also occur spontaneously as a self-protective response to emotional trauma. Deeply distressing experiences, such as rape, or violence on the battlefield, can trigger symptoms that have no apparent organic cause, including paralysis, amnesia, involuntary movements and hallucinations. In the past a person presenting with such dissociative symptoms would be labelled 'hysterical', though nowadays they would probably be diagnosed as suffering from a 'conversion disorder'.

There are some fascinating parallels between conversion disorders and hypnosis.[16] People who are prone to dissociative symptoms or who have experienced emotional trauma during their childhood tend to be easier to hypnotize, for example. Psychiatrists have long believed that the brain uses dissociation to protect itself from profoundly distressing experiences, perhaps creating a feeling of unreality or detachment so the person feels as though they were watching themselves in a film. One theory is that by developing the ability to divert conscious attention from interoceptive signals such as heartbeat or breathing rate, patients damp down their physiological responses and in so doing blunt their emotions. They use absorption to numb their suffering.

In one study, victims of rape who reported a high degree of dissociation during the attack (such as feeling strangely detached from what was happening to them) were found to have different physiological responses compared with victims who experienced less dissociation. When they discussed the rape with a researcher two weeks after it happened, their heart rate and skin conductance (a measure of sweating) were lower. The same suppression of physiological responses has been found among adolescent victims of child abuse or neglect and people involved in road traffic accidents who reported strong feelings of dissociation. Psychologists note that this isn't a healthy sign if it means people fail to process or come to terms with traumatic events because they are suppressing their physiological responses. During and shortly after a traumatic experience, they argue, dissociation may have the unfortunate effect of prolonging the symptoms of post-traumatic stress.[17]

Much of this chapter has been about the physiological changes that can be wrought by changes in expectation, and the otherwise impossible physical feats that can be performed. But the power of expectation doesn't by any means end with the physiological and the physical: it also reaches deep into the realms of spirituality. Spiritual or mystical states of consciousness have repeatedly been linked to activity in the temporal lobes of the brain. People with temporal lobe epilepsy, for example, can have profound experiences of this kind during a seizure. It therefore made sense that when researchers led by Michael Persinger at Laurentian University in Ontario used weak magnetic fields to stimulate the temporal lobes, volunteers often reported seeing visions, hearing voices or sensing the presence of an invisible sentient being. Many of these experiments were carried out in the nineties and the findings were widely reported in the media, including the BBC, CNN, the Discovery Channel and popular science magazines. Somewhere along the line, a journalist dubbed the headgear that was used to deliver the weak magnetic fields 'the God Helmet'. The original device was developed by Persinger's colleague Stanley Koren, a psychology professor at the University of British Columbia in Vancouver, from a modified snowmobile helmet, with solenoids placed over the wearer's temporal lobes.

You can buy a God Helmet online for around $650 (£500), which sounds like a lot of money until you realize that this equipment is so awesome it even works when it's switched off. Researchers at Uppsala University and Lund University in Sweden discovered this when they were monitoring the equipment's effects on undergraduates. These included feeling the presence of a sentient being, visions, hearing voices and a sense of oneness with everything or insights into ultimate reality. But it made no difference whether the God Helmet was switched on or off. Rather, individuals' experience of these phenomena was entirely dependent on factors such as whether they had pre-existing signs of temporal lobe anomalies, and their personality, for example whether they subscribed to New Age ideas and their capacity for absorption – which, as we've seen, underlies hypnotic suggestibility.[18]

It turned out that the God Helmet was not sparking spiritual experiences by magnetically stimulating the brain, but as a highly effective prop for facilitating hypnotic suggestion. It was the technological equivalent of the hypnotist's fob watch swinging from side to side before the patient's eyes. Since this discovery, sham brain-stimulation devices have proved invaluable as a powerful way to induce mystical experiences in the lab so that their signature in the brain and their effect on behaviour can be investigated.[19]

A cynic might cite such experiments as evidence these phenomena are 'all in the mind'. They may be right, though one might just as plausibly dismiss our five senses, our thoughts and emotions as 'all in the mind' too, since all conscious experience can be induced through suggestion alone. A wise sceptic should view findings like these as a further reminder to keep asking questions – 'Am I dreaming?', 'Am I being hypnotized?', 'Is this real?' – especially whenever charismatic, influential individuals, such as spin doctors, religious leaders and politicians are involved.

The findings in this chapter support the view that our brains are machines for inferring the causes of sensory data and that top-down predictions are what drive us. As a consequence, our behaviour and conscious experiences are easily manipulated by external suggestions. There are also times when our brains get things seriously wrong as a result of inbuilt susceptibilities in their hardware or the ingestion of psychoactive chemicals. The next two chapters explore the intriguing relationship between psychedelics and the delusions and hallucinations that characterize psychosis. This sets the scene for an exploration of the considerable healing potential of altered states in the final three chapters: their remarkable ability to restore balance and sanity.

## *How to Hypnotize Yourself*

In self-hypnosis, sometimes known as auto-hypnosis, you harness the power of your imagination to create a safe space in which to transform the deep cognitive models that govern your thoughts, emotions and behaviour. These models include conditioned fears and expectations that can worsen physical pain and that underlie anxieties about scenarios such as meeting new people, job interviews, exams and public speaking. Self-hypnosis provides an opportunity to unlearn this unhelpful conditioning. It can also help you kick bad habits such as smoking, overeating or drinking too much.[20,21]

In self-hypnosis, as in hypnosis led by a therapist, you remain in total control of what happens and can return to ordinary consciousness any time you like. In common with hypnotherapy, there is an induction phase designed to relax you and open your mind to change, followed by a suggestion phase in which positive affirmations are made about your goals. The only difference is that in self-hypnosis you play the roles of both hypnotist and patient: you dissociate yourself into a wise therapist and their receptive subject.

As with all forms of hypnosis, the following factors will maximize its efficacy:

- *Motivation*. Bring an open-hearted, trusting attitude to the practice. The impact of self-hypnosis will deepen with regular repetition, so try to set aside twenty or thirty minutes every day to achieve your chosen goals.

- *Concentration*. Ensure you won't be disturbed. Find somewhere quiet and turn off your phone.
- *Vividness*. Deploy all five senses in the scenes you conjure up: their associated tastes, smells, colours, sounds and textures. Make your imaginings as detailed as possible.
- *Positivity*. When you express your aspirations use only clear, positive statements in the present tense. For example, don't say: 'I will try not to mumble or stumble over my words during my interview', rather: 'I speak my mind with quiet confidence throughout the interview.' Repeat each statement three or four times with conviction.

Prepare a rough script outlining the imagery you will use and the affirmations for change you will make. The script provided below (Step 5 onwards) can be adapted according to your needs and imagination, or you could write your own. If you don't think you'll be able to remember it all, pre-record it so you can play it back to yourself.

1. Sit in a comfortable chair with your hands resting on your knees and your legs uncrossed with the soles of your feet flat on the floor. Keeping your head level, gaze at a point on the wall opposite.
2. As you do this focus on your breathing until it has become slow and steady. Empty your mind.

Whenever thoughts arise, just let them go.

3. After a few minutes, you will feel your eyelids become increasingly heavy, as though you are falling asleep. When you're ready, let them close.
4. Take a few slow, deep breaths. There's no hurry. Let your body sink into the chair, loose and limp as a rag doll.
5. Picture yourself standing on a harbour wall on a warm, sunny day looking out over a silver sea dotted with white sails under an immense blue sky. Your hands are resting on the parapet, and you can feel the sun-warmed, rough texture of the stone beneath your fingertips. Spend some time listening to the waves splashing against the foot of the wall below and the cry of seagulls above. Enjoy the warm sea breeze on your face and the smells of seaweed and salt spray.
6. There are five stone steps leading down to a quay where a small boat is moored in the calm waters enclosed by the harbour wall. Take the steps down to the quay, slowly, one by one, feeling the hard, even stones through the soles of your feet. Count them down in your head as you descend: 5–4–3–2–1. With each step you feel more relaxed, more contented, safer and more at ease.
7. When you're standing on the quay, you notice a wooden treasure chest in the bows of the boat. The key is in your pocket. Unlock the chest, lift the lid and take a look inside. Whatever you

ordered is in there: a new job, a qualification, confidence, health, happiness. There are objects symbolizing each – a pay cheque, a certificate, a trophy, a beautiful, sweet-smelling rose, whatever you like. Have a good look at the contents of the chest.

8. Now take out the objects one by one and put them in the rucksack you've brought with you, and as you do so, say out loud with conviction and self-assurance the affirmations you've prepared for each, such as: 'I have the skills to do this job. I am perfectly qualified for this work. I make my case to the interviewer with quiet, friendly confidence.' Repeat two more times, picturing yourself achieving your goal.
9. Once the chest is empty you can offload your anxieties, worries or bad habits into it. Perhaps they are represented by little grey packages labelled with luggage tags, or an ugly cigarette packet plastered with health warnings. You take them all out of your pockets and place them one at a time in the chest. As you do so, as appropriate, say something like, 'I am letting my anxieties and worries go' or 'I am relinquishing my bad habits.' When all the objects are in there, close the lid and lock the chest.
10. Untie the boat and give it a gentle push. Watch it drift towards the harbour entrance with the tide, out through the gap and onto the open sea. 'I

am waving my anxieties and bad habits goodbye,' you say, waving, and as the boat recedes into the distance you feel lighter and lighter. 'I am waving my anxieties and bad habits goodbye.'

11. When you are ready, climb the steps back up to the harbour wall, feeling more awake with every step. Count them up as you ascend: 1–2–3–4–5. When you reach the top you are energized and alert. In your own time, open your eyes and spend a few moments relishing your renewed sense of optimism and purpose.

# 5

# *Wonder Child*

A chemist called Albert Hofmann was hard at work in his lab in Basel, Switzerland, one spring afternoon in 1943 when he started to feel a little odd. Writing a report for his boss the next day, he described the sensations as 'a remarkable restlessness, combined with a slight dizziness'. Unable to concentrate, he left work early. Back home, lying down and closing his eyes, he hallucinated 'an uninterrupted stream of fantastic pictures, extraordinary shapes with intense, kaleidoscopic play of colors'.

Everyday reality was restored a few hours later. What on earth had happened? He figured that despite his usual meticulous precautions, a drop of the solution he had been crystallizing that afternoon might have touched his fingertips and been absorbed through his skin. Five years earlier, synthesizing this particular substance for the first time, he hoped it would be useful as a respiratory and circulatory stimulant, but when the animal test results came back they were disappointing: nothing much to report, certainly none of the hoped-for physiological effects – just a casual observation that the animals became 'restless'. Five years on, a 'peculiar presentiment' that his colleagues might have missed something had driven him to synthesize a few hundredths of a gram more. But those hallucinations! Where had they come from?

Hofmann resolved to settle the matter once and for all. If a minuscule amount absorbed through his skin really was the culprit, he reasoned, it must be strong stuff. So a few days after his initial experience he swallowed what he assumed to be a vanishingly small dose – 250 micrograms, lighter than a grain of table salt – and waited to see what would happen. Forty minutes later it was only with great effort he managed to scrawl in his lab journal: 'dizziness, anxiety, visual distortions, symptoms of paralysis, desire to laugh...' Struggling to speak intelligibly, he begged his assistant to escort him home. Wartime restrictions were in force on the use of cars, so they had to ride by bicycle through the crowded streets of Basel. 'On the way home, my condition began to assume threatening forms. Everything in my field of vision wavered and was distorted as if seen in a curved mirror.'

Somehow arriving home safely, in a final moment of lucidity the chemist sent his assistant next door for milk, which at the time was commonly believed to be a general antidote for poisoning. He felt awful. Everyone and everything was assuming increasingly frightening forms and, to his horror, when his neighbour arrived with the milk she was nothing like the woman he knew but 'a malevolent, insidious witch with a colored mask'.

If anything, the changes in his inner being were even more disturbing. Recalling the experience years later, Hofmann wrote: 'I was seized by the dreadful fear of going insane. I was taken to another world, another place, another time. My body seemed to be without sensation, lifeless, strange. Was I dying? Was this the transition?' His assistant called a doctor, who found nothing physically wrong: pulse, blood pressure and breathing were all normal. Nonetheless, the doctor watched over him late into the night and, as the hours passed, the chemist slowly regained his mental faculties. He even began to enjoy the fireworks going off inside his head. 'Kaleidoscopic, fantastic images surged in on me, alternating, variegated, opening and then closing themselves in circles and spirals, exploding in colored fountains, rearranging and hybridizing themselves in constant flux.' Exhausted, he finally slept, awakening the next morning feeling refreshed and with a clear head:

> A sensation of well-being and renewed life flowed through me. Breakfast tasted delicious and gave me extraordinary pleasure. When I later walked out into the garden, in which the sun shone now after a spring rain, everything glistened and sparkled in a fresh light. The world was as if newly created. All my senses vibrated in a condition of highest sensitivity, which persisted for the entire day.[1]

The substance Hofmann had brought sparkling and screaming into the world was lysergic acid diethylamide, better known as LSD. No other chemical synthesized before or since triggers such profound changes in consciousness at doses so low they are measured in millionths of a gram. What exactly is it? In common with many of our most potent drugs, we have nature to thank for LSD. This hell-raising rock star of the chemical world is an alkaloid with humble origins, a derivative of a toxic fungal infestation of rye known as ergot. All ergot alkaloids, both the natural ones and the many variants cooked up in chemistry labs, comprise lysergic acid bonded to another molecule. When Hofmann joined the Swiss drugs company Sandoz in 1929, chemists had been tinkering with these substances for more than a decade because of their ability to constrict blood vessels. The first to be isolated in its pure form, ergotamine, is still in use today as a treatment for migraines and to stop bleeding after childbirth.

Hofmann began synthesizing ergot alkaloids in 1935, combining lysergic acid with a class of organic chemicals known as amines. Three years on, among the twenty-four candidates he had created, one had found success as Methergine, a drug for inducing labour and staunching bleeding, but the twenty-fifth child born into this huge family would go on to outshine all the others. Of course things didn't turn out too well for LSD and, after a remarkable adolescence, the prodigy became – in Hofmann's own words – his 'problem child'.

Some view LSD's birth as a happy accident, but if it weren't for Hofmann's skills as a chemist, his scientific instincts and intense, frankly reckless curiosity, this exceptional substance might have languished unnoticed in a vial on a dusty shelf at Sandoz. As it was, over the next two decades LSD built a reputation among artists as a chemical key that could unlock their creativity. Among psychiatrists, it was quickly recognized as a powerful aid to psychoanalysis: low doses made patients less guarded during sessions and seemed to throw open a window into their subconscious, whereas doses as high as that taken by Hofmann mimicked the disturbing effects of psychosis. Brave therapists would sometimes take LSD to gain insights into their schizophrenic patients' mental state. Others hoped it would lead to a new treatment for the condition. Moderate doses seemed to help alcoholics overcome their addiction and ease the pain and anxiety of life-threatening illnesses.

It's worth pointing out that these early forays into psychedelic therapy were not the whims of quacks or crackpots on the medical fringes. Between 1950 and 1965 more than a thousand clinical research papers about LSD were published, between them reviewing the treatment of some 40,000 patients. Dozens of books were written about the new therapy, specialist treatment centres were established and six international conferences were held.[2]

By the early sixties, however, LSD or 'acid' had seeped from the tightly regulated environment of laboratories and clinics into the happy chaos of popular culture. Hofmann was bemused by this turn of events – he never dreamed anyone would want to experience the strange, often nightmarish effects of LSD just for the fun of it. His prodigy's transformation from a serious therapeutic agent into the counterculture's drug of choice was largely the result of proselytizing by two academic psychologists, Timothy Leary and Richard Alpert (later known as Ram Dass), who were booted out of Harvard in 1963 after their freewheeling research projects with LSD and psilocybin (found in magic mushrooms and truffles) gained a reputation on campus for being more akin to psychedelic parties than scientific studies. In the eyes of the science faculty, their worst sin was that they had fatally compromised their objectivity by taking the drugs themselves during experiments.

Leary would go on to exhort American youth to 'turn on, tune in, drop out'. He explained to a TV journalist:

> We teach the science and art of ecstasy. We teach people how to turn on, or how to go out of their minds... to get beyond routine ways of thinking and acting and experiencing. We often say that we're teaching people how to use their head. The point is that in order to use your head you have to go out of your mind.

Americans were living in an 'insane asylum', he told another interviewer, 'hung up' on materialism, power and war-making. By the mid-sixties LSD was freely, legally available and in widespread use among young people. Leary expressed his ambition that within ten years 20–30 million Americans would be using the drug regularly.[3]

Uncle Sam had other ideas – for one thing there was a war to be fought in Vietnam. Richard Nixon dubbed Leary 'the most dangerous man in America' and so the law began to close in on Hofmann's problem child as government agencies imposed increasingly strict regulations on the scientific and therapeutic uses of psychedelics. Possession was banned in the US in 1968 and medicinal use outlawed under the 1970 Controlled Substances Act, which listed the drugs in Schedule I – alongside heroin and methamphetamine – as having high potential for abuse and 'no currently accepted medical use'. To this day in the US, LSD is officially considered more dangerous and less medically useful than cocaine, which is listed in Schedule II. The prohibition on psychedelics went global in 1976 when the UN Convention on Psychotropic Substances (signed five years earlier) came into force, rendering further research prohibitively expensive and bureaucratically fraught. Its reputation shredded, to study LSD or even express an interest in its potential medical uses became an act of professional suicide.

It goes almost without saying that the legal clampdowns of the sixties and seventies have done little to curb the availability of acid and other psychedelic drugs. In 2010, an estimated 32 million Americans had taken a psychedelic during their lifetime and 23 million had taken LSD.[4] But for four decades not a single clinical study saw the light of day. Against all the odds, however, the past few years have witnessed a revival in scientific interest. In the US, the first federally approved clinical research project involving LSD since the seventies was published in 2014: a pilot study by American and Swiss psychiatrists whose results suggested that, at low doses and in combination with psychotherapy, the drug can alleviate anxiety in patients with terminal illnesses.[5] Larger studies published in 2016 found that psilocybin can help lift the depression associated with a life-threatening cancer diagnosis, with improvements lasting at least six months.[6,7] Early clinical trials have also hinted that psychedelics – professionally administered in a controlled environment – have roles to play in treating major depression, alcoholism and helping people quit smoking.[8–10] Intriguingly, two recent studies suggest that both ayahuasca (the DMT-containing psychedelic tea) and psilocybin can boost people's mindfulness abilities, at least temporarily, to levels usually only achieved after years of meditation practice.[11,12]

I explore some of these tantalizing results in more detail in the next chapter, but first it's worth delving into the molecular biology and neuroscience of these mind-boggling chemicals. How do they precipitate their startling effects – the ugly and the beautiful, the terrifying and the transcendent – and what can they tell us about ordinary consciousness?

Much of the credit for the renaissance in psychedelic research must go to Amanda Feilding, a rebellious English aristocrat and close friend of Hofmann in the final years of his life. In 1998 Feilding set up the Beckley Foundation to initiate scientific studies into consciousness-altering drugs and campaign for evidence-based reform of global drugs policy. In the sixties, Feilding experimented freely with her own consciousness, taking LSD while it was still legal, smoking pot and, on one occasion, trepanning herself – drilling a small, neat hole in her skull while her partner filmed the bloody procedure (none the worse for the supposedly consciousness-expanding procedure, that evening she wrapped a brightly coloured scarf around her head and went to a party). These days the consciousness-altering research that her charitable trust supports is thoroughly mainstream, involving collaborations with scientists at world-class institutions including Imperial College London, Johns Hopkins University School of Medicine in the US and the Sant Pau Biomedical Research Institute in Barcelona, Spain.

In April 2016, Feilding and Imperial published the world's first images of the human brain on LSD.[13] The research was led by Robin Carhart-Harris and provides striking insights into the changes wrought by the chemical that Hofmann first synthesized eight decades earlier. It helps to explain the experience in purely scientific terms: the spectacular hallucinations, the spiritual insights, the terrifying transformations.

I met Feilding for the first time shortly before the results were published, interviewing the energetic, charismatic seventy-three-year-old at her foundation's headquarters at Beckley Park, a Tudor hunting lodge with three towers and three moats near Oxford where she has lived all her life. Her formal title is the Countess of Wemyss and March, though practically everyone she knows calls her Amanda. As we sat in her office sipping Earl Grey tea, I asked her what tripping on LSD felt like. She enthused:

> It's about being more yourself, more sensitive, more attuned to beauty and music. And the wonderful thing about LSD that people often don't know is how dose-selective it is. It can go anything from a microdose so you hardly notice it – you just feel a bit of sparkle, a bit more energy – and then there's a medium dose where you see beauty more and you hear more, up to a much bigger dose where there's ego dissolution. It's a whole spectrum of changes.

In the sixties some of her friends were swallowing as much as 250 micrograms at a time – the dose that sent Hofmann's brain spinning off on its alarming trip. At the more conservative end of the scale, she said, 15 micrograms simply brightened one's mood. 'So in a sense, the art of these substances is the art of getting the dose right and controlling your state of consciousness within the dose.' Crucially important in this regard is what Leary called 'set and setting' – your mindset and the environment where you take the drug. Noisy, chaotic surroundings where strangers come and go can be particularly disturbing for someone who has taken a high dose of LSD, making it an unwise choice as a party drug.

For the purposes of the Beckley/Imperial brain-imaging study, twenty volunteers were given a moderate dose of 75 micrograms. Each healthy subject had his or her brain imaged on two occasions at least two weeks apart, receiving an intravenous shot of LSD at one session and a placebo at the other. To cross-check the findings, three different imaging techniques were used: BOLD (blood oxygen level dependent) MRI, which indirectly maps the brain's energy expenditure; another type of MRI called 'arterial spin labelling' that measures blood flow; and magnetoencephalography (MEG), which records the amplitude and frequency of brainwaves. During imaging the volunteers pressed buttons to indicate the intensity of their visual hallucinations and any sense of ego dissolution (loss of subjective selfhood), and after each scan they filled out a standard questionnaire designed to assess altered states of consciousness.

On LSD, it was as if top-down control of the brain's operations had been suspended, precipitating a breakdown in connectivity within its networks and increased chatter between regions that are not usually on talking terms. The strength of alpha waves – synchronous electrical pulses oscillating at 8–15 Hz that orchestrate top-down brain activity – decreased, which may explain the breakdown in normal patterns of connectivity. In ordinary consciousness, alpha waves are thought to help focus attention by suppressing activity in the swathes of nerves that aren't needed for that particular task.[14]

The neural origins of Hofmann's 'uninterrupted stream of fantastic pictures, extraordinary shapes with intense, kaleidoscopic play of colors' were revealed. Despite subjects keeping their eyes tightly closed, blood flow was boosted in their primary visual cortex and its connectivity with other regions of the brain was strengthened. These stronger links may also help to explain LSD's ability to evoke synaesthesia, in which people report a crossover between usually discrete senses. Sounds may trigger visual effects, for example. In one of my own psychedelic adventures – on psilocybin-containing truffles rather than LSD – the sound patterns of a particular piece of classical music powerfully evoked the sensation of floating among the rafters of some lofty, cathedral-like space.

The magnitude of changes in blood flow and alpha waves in the brains of volunteers on LSD correlated closely with the extent of their visual hallucinations. Meanwhile, changes in their consciousness – typified by a sense of their ego dissolving – were associated with decreased connectivity between components of one of the brain's most important networks, known as the default mode network or DMN. Like the visual hallucinations, this breakdown seemed to be caused by the waning power of electrical oscillations that usually coordinate the activity of far-flung brain regions. There was a particularly steep reduction in the strength of delta waves (pulsing at 1–4 Hz) and alpha waves originating in a highly connected hub within the DMN called the posterior cingulate cortex (PCC).

The DMN has been a hot research topic among neuroscientists ever since it was discovered in 2001. Its nodes become more active in the brains of subjects lying flat on their backs in scanners when they're not given anything in particular to do, which is why the network was initially dismissed as a kind of baseline, low-power or 'standby' mode of operation. But it later transpired that the network is surprisingly power-hungry. Energy consumption in its principal hub – the PCC – is 40 per cent higher than the average for the rest of the brain. To gobble up this much valuable energy, the network must serve some vital survival function. One clue is that the DMN developed rapidly in the course of our species' recent evolution, during the 6 million or so years since our line split from that of the other great apes. Another is that its constituent regions are more loosely linked up in infants and children, only reaching full connectivity in adulthood.[15]

The network is now known to draw upon autobiographical memories in order to project our minds into the future and the past, generating 'self-referential' thoughts and our sense of selfhood. It not only creates our sense of autobiographical selfhood or ego, but also allows us to simulate the perspectives of other selves in our social world (known as 'mentalizing' or 'theory of mind'). It will be obvious to any parent that the social sophistication and sensitivity of their child steadily increases as she or he grows up, reaching an occasionally uncomfortable pitch in adolescence. Primatologists assure us that the mentalizing abilities of human adults easily exceed those of our closest evolutionary cousins, chimpanzees, which have less insight into the minds of other chimps. Perhaps as a result, they band together in much smaller, simpler social groups compared with those of our own, more socially savvy species.[16]

I asked Feilding to explain in everyday terms what happens to the DMN in people who have taken LSD. 'The default mode network is the wretched, dominant, bossy network,' she said with feeling. 'It's the top-down mechanism that maintains control over brain activity, rather like a government does over its citizens.' The kind of control it exerts is determined by our conditioning – the established norms about how to behave, what to aspire to, what to think and what to say that we have absorbed over the course of a lifetime. So the DMN may be partly responsible for the feelings of guilt and judgement we experience when either we or other people transgress these norms. LSD disrupts its activity. As Feilding explained:

> When the default network is knocked out, it's like the conductor of an orchestra no longer being there, so all the different instruments can start to make their own sound and do their own thing. They become more united with everything else and with the external world. That opens up new possibilities of connectivity and 'Aha!' moments of realization.

Nonetheless, like the conductor of an orchestra or a central government, the DMN serves an invaluable purpose. What LSD and other classic psychedelics help to reveal is how a healthy brain, during ordinary consciousness, regulates the activity of the billions of nerve cells that are its citizens, maintaining a judicious balance between rigid order and anarchy. Carhart-Harris and his psychedelic research lab at Imperial have proposed that, under the influence of these drugs, the brain reverts to a more disordered, 'primary' state of consciousness that has features in common with childhood, dreaming sleep, the early phase of psychosis and the dreamy state that precedes a fit in temporal lobe epilepsy. All are characterized by a mode of thinking less constrained by the continual sensory and cognitive reality-checking of everyday adult consciousness. In these states people exhibit what psychologists call 'magical thinking', such as belief in supernatural phenomena, wishful thinking, fantasies and paranoia.

Another feature that several primary states of consciousness share is visual hallucinations, which clearly characterize dreams and the psychedelic state but can also be experienced in early psychosis and in the moments leading up to a fit in temporal lobe epilepsy. The visions of people in the throes of religious ecstasy or deep in meditation are also beginning to look suspiciously like further examples of disordered, primary states of consciousness. Ego dissolution – even to the extent of feeling 'at one with the universe' – is another common thread running through these experiences, with the possible exception of dreams and childhood.

Feilding and the Imperial team have suggested that the conductor or organizing principle minimizing disorder in the brain and preventing us all from experiencing these mystical effects all the time is the DMN, the neural substrate of our sense of autobiographical selfhood or ego. They call their proposal the 'entropic brain hypothesis'.[17] Entropy is a measure of the disorder or unpredictability of a physical, chemical or biological system. Many systems that comprise large numbers of individual components, whether it's the molecules in a gas, the grains of sand in a dune or a flock of starlings, are known to have 'emergent properties' as a result of their innate tendency to coalesce into more orderly structures. They are also vulnerable to abrupt changes – like avalanches in a sand dune – that occur at the precariously balanced, 'critical' boundary between organized and disorganized phases.

The entropic brain hypothesis builds upon the idea that the brain, in common with many living systems, displays *self-organized* criticality, whereby extraordinary complexity can emerge from simple local interactions between nerve cells. The brain essentially tunes itself (or 'self-organizes') to occupy the sweet spot between order and disorder where psychological phenomena such as selfhood can emerge. Feilding and her colleagues propose that in secondary consciousness (regular adult waking consciousness), the DMN holds billions of nerve cells in a relatively stable state just close enough to criticality, with our highly developed sense of 'self' and 'other' arising as an emergent property essential for a highly social species such as ours. Like the conductor of an orchestra, the default mode imposes order on a multitude of individual instruments – the massed ranks of nerve cells. The result is ego: the ambient music playing in the background of ordinary adult consciousness.

By contrast, in primary states of consciousness, such as early psychosis, temporal lobe epilepsy and the psychedelic state, the DMN is forced to relinquish control, raising entropy in the brain and pushing it perilously closer to criticality, with unpredictable, quite possibly discordant consequences for perception and cognition (see Figure 5). The default mode, as Feilding put it to me, does tend to be 'bossy', but in any biological system – not just the brain – a delicate balance must be struck between order and flexibility. One way to think about this is to picture the mental landscape as a flexible rubber sheet with a marble rolling about on it. Our cognitive models of the world are represented by indentations or troughs in the sheet into which the marble can roll and come to rest. Statisticians call these troughs 'basins of attraction' and the stable states at the bottom of them 'attractors'. If the basins of attraction within our mindscape are too deep (low entropy), the marble will get stuck in one of them, but if they're too shallow (high entropy) it will roll about more or less at random. The first state is rigidly inflexible, the second mindlessly chaotic.

Getting the balance right between these undesirable extremes presents a dilemma for a biological information-processing system like the brain because, while it's wired to learn about its environment in increasingly fine detail, it also needs to retain enough flexibility to adapt when the environment changes. A highly accurate but hidebound model of its world (a low-entropy landscape pitted with deep troughs) could spell death in times of famine, for instance, when more creative 'out of the box' thinking is required to discover new food sources and avoid starvation.

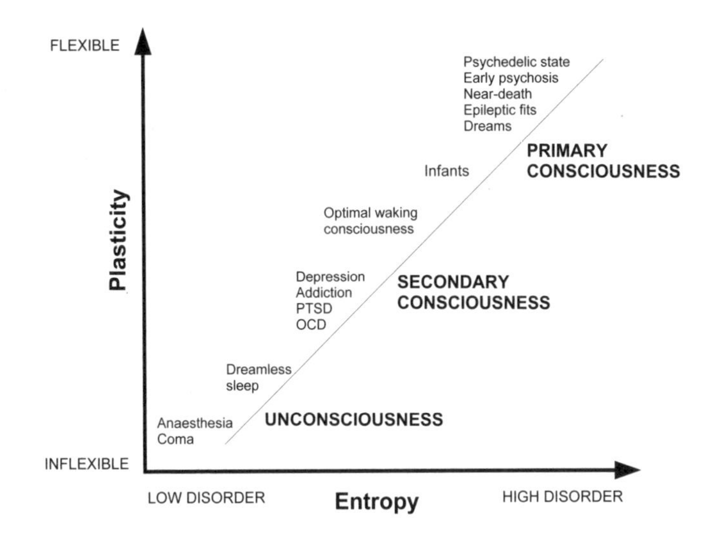


Figure 5: *The entropic brain hypothesis proposes that there's a delicate balance between flexibility and disorder in a healthy brain*

As we've already seen in this context, our dreams are essential for overnight maintenance, restoring the balance between rigidity and flexibility after a day's learning experiences. In Chapter 2, I touched upon the proposal by Karl Friston, Allan Hobson and Charles Hong that dreams streamline or optimize our brains' models of the world. The idea is that while we're awake the models become increasingly accurate but less and less efficient. Like overly complex scientific hypotheses that draw upon lots of different parameters to explain the available data, this makes them cumbersome and less widely applicable. When the 'marble' of consciousness rolls into one of these deep basins of attraction it's more likely to get stuck there, resulting in the kind of inflexible thinking and behaviour that characterize addiction, obsessive compulsive disorder (OCD), post-traumatic stress disorder (PTSD) and depression. Only when our senses and muscles are offline during sleep, the scientists argue, can the brain's prediction processing hierarchy regain some of its former flexibility by 'rebooting', pruning the profuse synaptic connections that have sprouted during the day. According to Friston, Hobson and Hong, the chaotic, quasi-psychotic fantasy worlds of our dreams that the brain spits out while this is happening are a diverting but ultimately purposeless side effect.[18]

Other primary states of consciousness, including LSD trips, may fulfil much the same role, temporarily boosting entropy in the brain – nudging it closer to criticality – with unpredictable and disorganized, but potentially therapeutic results. A reboot may be just what's needed to shake out the rigid patterns of thought and behaviour that characterize many mental illnesses.

So psychedelics' restorative, ego-busting effects are largely down to their ability to play havoc in the DMN. Both psilocybin and LSD are now known to reduce blood flow in the network and quieten chatter among its component regions (their 'functional connectivity' in neuroscience parlance). Like all the classic psychedelics, including mescaline and DMT, their mind-altering properties have been traced to a particular type of serotonin receptor found in the membranes of nerve cells in the deepest layer of the cortex where the brain's top-down predictions originate. Specifically, the receptor is found in the 'post-synaptic' membranes of these neurons where signalling molecules from other nerve cells dock after crossing the narrow channel that divides them (see

Figure 6). Known as the serotonin 2A receptor, it is particularly abundant in the PCC – the principal hub of the DMN. All the classic psychedelics bind to it, like keys fitting snugly in a lock, as a result of their structural similarities with the receptor's usual binding fellow, the brain hormone serotonin. LSD has a particularly high affinity for the receptor (it latches on and won't let go), which helps explain the drug's extraordinary potency at very low doses. When serotonin or a psychedelic molecule bind to the receptor they 'depolarize' the membrane, reducing the voltage across it. Like gently squeezing the trigger of a gun, this increases the likelihood that the nerve cell will fire.

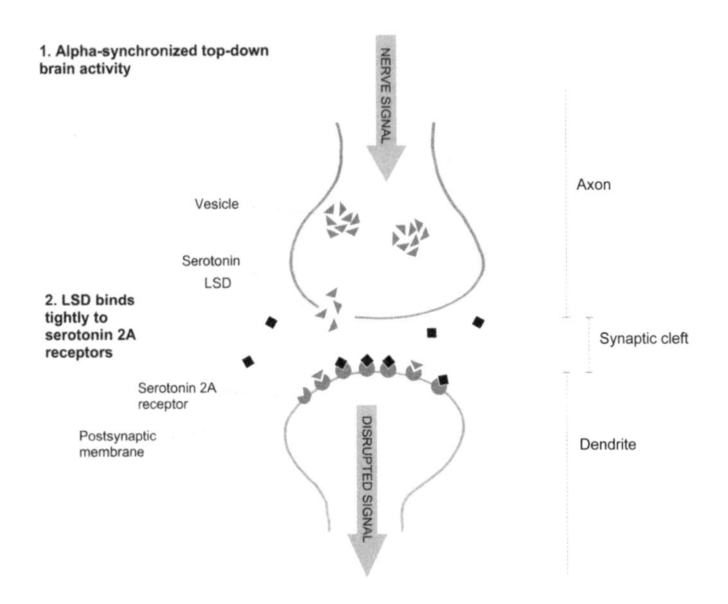


Figure 6: *When LSD binds to serotonin 2A receptors this disrupts the ability of nerve cells to fire in synchrony*

There remains some uncertainty at this point in the molecular tale, because serotonin 2A receptors pepper the synapses of two very different types of nerve cell in the PCC, one stimulatory and one inhibitory. The Imperial researchers have yet to nail the precise sequence of events, but they can be confident that when psychedelic molecules bind to serotonin 2A receptors this stymies the PCC's ability to generate the orderly alpha waves that marshal vast assemblages of nerves across the cortex into firing in synchrony. Supporting this view is the fact that both LSD and psilocybin reduce the intensity of alpha waves originating in the PCC. To adapt Feilding's metaphor, it's as if some prankster has hidden the conductor's baton. Crucially, in people who have taken psilocybin, the magnitude of this decrease in alpha waves correlates with their reports of ego dissolution and increased tendency for magical thinking.

If the entropic brain hypothesis is true, by impeding the work of the DMN, psychedelics regress the brain to a more primitive, child-like or chaotic form of consciousness less dominated by the ego and less fastidious about reality-checking. In ordinary waking consciousness, one of the DMN's most important tasks may be to keep a lid on the activity of the temporal lobes where a lifetime of memories have been encoded and stored. So when connectivity between the network and the temporal cortex is lost as a result of weakened alpha oscillations, the lid of this Pandora's box is blown off, letting loose a chaotic stream of memories that the visual cortex faithfully displays to the mind's eye as the visions of early psychosis, dreams, temporal lobe epilepsy and the hallucinogenic state. This ties in nicely with prediction error theory, which views the temporal lobes as the core of the brain's processing hierarchy, the source of its most abstract inferences about the world.

In a nutshell, the disruption of DMN activity that ensues after psychedelic molecules bind to serotonin 2A receptors in the PCC is probably responsible for most of the fireworks, spiritual revelations, smashed egos and emotional insights that people who have taken them report. But the implications don't end there because, intriguingly, the stories of several other altered states of consciousness – including hypnosis, trance, meditation and early psychosis – also converge here at the serotonin 2A receptor.

The theme that they all have in common is absorption. As we saw in the previous chapter, the capacity to focus attention to the exclusion of almost everything else underpins people's susceptibility to hypnotic suggestion (see page 94). Only by tightly constricting the focus of information processing in the brain can faulty models of reality be sustained in the face of conflicting evidence. Individuals' tendency to become highly absorbed in an activity is a stable personality trait bound up with other characteristics, such as vivid imagination, synaesthesia, a tendency to have paranormal experiences and intense emotional responses to beauty in nature and the arts – not unlike a mild psychedelic trip. Psychologists often measure the trait using a questionnaire called the Tellegen Absorption Scale which includes statements such as 'I can be deeply moved by a sunset', 'When I listen to music I can get so caught up in it that I don't notice anything else' and 'Sometimes I feel and experience things as I did when I was a child.' When they discovered the link between hypnotizability and absorption in 1974, the American psychologists Auke Tellegen and Gilbert Atkinson wrote that the 'objects of absorbed attention acquire an importance and intimacy that are normally reserved for the self and may, therefore, acquire a temporary self-like quality. These object identifications have mystical overtones.'[19]

Tellegen and Atkinson intuited more than forty years ago that the absorption necessary for hypnosis entails a temporary surrender of selfhood. It now turns out that absorption is also associated with susceptibility to the effects of psychedelic drugs. A recent study showed that the second most important factor determining the strength of a person's response to psilocybin, after dosage, is how prone they are to absorption.[20] Those who score high on the Tellegen scale are more likely, for example, to experience mystical and hallucinatory effects after taking the drug. What's more, people who score high on the scale are also more likely to have inherited a gene that makes a 'stickier' version of the serotonin 2A receptor that binds more tightly not only to serotonin but also to psychedelic molecules such as LSD and psilocybin. The very same genetic variant is associated with a person's susceptibility to the hallucinations, delusions and confused thoughts that characterize schizophrenia.[21]

It's worth noting here that dozens of genes have been linked to schizophrenia and that each, individually, is only weakly associated with the condition. No single gene or environmental factor is responsible for the illness. Rather, this linkage between psychosis, the serotonin 2A receptor and the personality trait of absorption adds to growing evidence that symptoms of mental illness exist on a continuum of intensity throughout the general population. It provides support for the increasingly mainstream view among psychiatrists that psychotic symptoms, such as delusions and hallucinations, are an extreme manifestation of universal attributes of human conscious experience.

According to a survey published in 2000, around 40 per cent of adults have knowingly had a daytime hallucination.[22] But, as we've already seen, the truth is all of us hallucinate all the time, albeit these hallucinations are constrained by the evidence of our senses. When these constraints are removed as a result of hearing or sight loss, when there's hardly any sensory data available, or the data is monotonously unchanging, people often hear and see the ghosts of real things. In Charles Bonnet syndrome, for example, which can affect older people who have lost or are losing their sight, the brain summons up highly convincing, detailed visions. In his book *Hallucinations* the neurologist Oliver Sacks recalled the case of an elderly lady in a nursing home who, despite having been completely blind for several years, began to see an endless parade of figures in colourful Eastern costumes going up and down stairs. Even though her blindness meant she couldn't see Sacks, he watched her eyes follow these phantoms as if they were actually there in the room with them.

Sacks went on to describe the visions that can console or torment people in solitary confinement, a phenomenon wryly termed 'the prisoner's cinema'. He wrote that in any sensory desert, not just prison cells, mirages of this kind are commonplace. Pilots, sailors and long-distance truckers, for example, are notoriously prone to seeing things that aren't really there. Sleep deprivation, as we've seen, can be a major contributory factor.[23]

Until comparatively recently, scientists took it for granted that what we see, hear, taste and touch is like a mental theatre set being continually assembled, dismantled and reassembled inside our heads using the sensory evidence garnered by our eyes, ears, taste, stretch and touch receptors. But if this were true – even assuming our brains have the processing power needed for this prodigious feat – they would be constantly on the back foot, struggling to keep up in a rapidly changing world of dangers and opportunities. Instead, many neuroscientists now believe that brains keep one step ahead of the game by *predicting* what's happening, drawing upon the accumulated evidence of a lifetime of experiences. Only the prediction errors or 'surprises' reported back by our senses – a mere dribble of data compared with the flood of sensory information that would be needed to build an entire perceptual theatre set on the fly – are allowed to penetrate its predictive machinery to update our rolling models of the world.

Hallucination is inevitable given that all conscious perception arises from the brain's top-down predictions – its 'best guesses' about the hidden causes of stimuli. 'Perception is just hallucination that's constrained by sensory evidence,' Friston told me when I met him at his office in University College London to talk about dreams, psychosis and other altered states. 'Dreaming is just hallucination that's *not* constrained by it.' That people who are wide awake sometimes see and hear things that aren't really there is an inescapable consequence of how perception works. Seen in this light, Friston said, 'hallucinations are perfectly functional. Lots of people who are not psychotic suffer from hallucinations.'

What is it about people under the influence of psychedelics, and those suffering from psychosis, that makes them unusually prone to hallucination? The answer may lie in the way the brain deals with uncertainty. Friston and other neuroscientists believe that the principal stumbling block for a brain trying to make its way in an uncertain world is that sensory signals are often 'noisy'. They are sparse, jumbled and ambiguous. Visual data, for example, is inherently unreliable in a dimly lit or fast-moving environment. This can lead to 'false positives' – seeing things that aren't there – and 'false negatives' – failing to notice or recognize something that actually is there. What's a brain to do? The prediction error theory that Friston helped to develop proposes that the brain filters sensory data as the information passes upwards through its processing hierarchy according to an estimation of reliability (the information's 'precision' in statistician-speak). This estimate is based not just on an assessment of how scattered the data points are (their 'variance') but also the brain's expectations about the trustworthiness of the source. These expectations are mostly unconscious and heavily dependent on past experience and the current context.

The molecular bouncers performing this filtering operation – blocking dodgy data and waving through trustworthy data – are known as 'neuromodulators'. They include noradrenaline, acetylcholine, dopamine and serotonin and work by amplifying the transmission of prediction error signals at synapses. We first encountered them in Chapter 2 in the context of dreams, the idea being that their relative levels determine whether we're wide awake, in dreamless sleep or dreaming. In waking consciousness neuromodulators regulate the relative influence of top-down predictions and bottom-up prediction errors according to the estimated reliability of these two streams of information. In a particularly 'noisy' environment, such as a dark alleyway at night, when the raw evidence of our senses can't be trusted, they will favour top-down expectations ('It's not safe here') to shape perception and behaviour. Under these circumstances, we will be primed to see menacing shapes hiding in the shadows. Walking in broad daylight down the same alleyway, the brain shifts the balance in favour of the more precise, less noisy sensory evidence that is now available to it.

Language is another prime example of how the brain uses its established cognitive models to impose order on noisy, ambiguous signals. When we hear someone speak our native tongue, multiple layers of structure and meaning are automatically overlaid on the sound data to create our conscious auditory experience. Display the same acoustic data graphically in a sonogram, however, and there will be little discernible structure. The transitions between consecutive words, for instance, which we hear so distinctly when somebody speaks will be nowhere to be seen.

In hearing, as in all perception, our prior expectations are key. This makes the occasional 'false positive' inevitable. People can be primed to hear the lyrics of a particular song within a meaningless blizzard of sound, for example. When Dutch psychologists played white noise to healthy, 'neurotypical' undergraduate students and asked if they could discern a faint trace of Bing Crosby's 'White Christmas', around a third were convinced that they could.[24]

A striking demonstration of the same phenomenon became an instant hit on Twitter in May 2018 when someone posted a video excerpt of a plastic toy depicting Brainstorm, a crab-like, big-brained alien from the children's animated TV series *Ben 10: Alien Force*. The transparent effigy pulses with green light when a button is pressed and you hear a voice say the word 'brainstorm' – provided 'brainstorm' is what you were *expecting* to hear. If you think 'green needle' before the button is pressed, you distinctly hear the words 'green needle'. You can also hear 'brain needle' and 'green storm' if those were the words you were thinking before the button was pressed. You get the unsettling impression that the toy is reading your mind. Perhaps the most surprising thing about this auditory illusion is that, apart from containing different consonants, 'needle' has two syllables and 'storm' has one. Interestingly, the light pulses three times as the voice speaks, which may help cue you to hear the extra syllable if that's what you were expecting. If you don't believe that perception is nine-tenths hallucination, check out the illusion for yourself at https://bit.ly/2IPgrh3.

It's a small hop from the auditory hallucinations everyone experiences to the auditory hallucinations of psychosis. A friend who has schizophrenia tells me there are times when he hears his neighbours talking about him through the wall that divides his house from theirs. If his condition deteriorates and we're walking down a crowded street he will often start to obsess about the derogatory comments he can hear passers-by making about him – comments that, of course, I can't hear.

Somewhere along the line, the delicate balance of credence between my friend's top-down cognitive predictions about the world and the bottom-up prediction errors reported back by his senses has been lost. Friston uses the analogy of browsing the Internet for news. Everyone can agree that the most surprising information – the stuff no one predicted – is the most newsworthy, but you are still left with the delicate task of assessing the reliability of its source. The information could be 'fake news' – propaganda or just another conspiracy theory. As we've seen, the solution that the brain appears to have evolved is that neuromodulators adjust the 'gain' or volume of the prediction error signal according to its estimated 'precision' or trustworthiness.

The hallucinations and delusions that characterize both psychosis and the psychedelic state seem to be the result of breakdowns in this process of assigning confidence to ascending sensory signals relative to descending predictions.[25] In somebody vulnerable to psychosis, their difficulties may start when inbuilt faults in their dopamine gatekeeping system give too much credence to noisy sensory data, causing visual and auditory hallucinations. Over time these false positives build increasingly inaccurate cognitive models of the world which, in turn, skew the brain's sensory expectations, setting in train a vicious cycle of hallucination, delusion and paranoia.

In a person who has ingested a psychedelic drug, it's the serotonin gatekeeping system that is disrupted, albeit temporarily. When psychedelics bind to serotonin 2A receptors in regions such as the PCC that sit high in the information-processing hierarchy, this shifts the balance of trust in favour of bottom-up sensory data relative to the usually dominant top-down predictions. This can allow noisy, unreliable sensory information to influence conscious perception. Friston explains:

> When you give psychedelics you break the ability of the PCC to discriminate between Fake News and Precise News. You can see how that would have a profound effect on its ability to orchestrate perceptual synthesis, attending to the wrong sorts of things even to the extent of attending to percepts that are not there.

In ordinary consciousness, alpha waves organize our top-down sensory predictions along well-established lines. Brain-imaging experiments with LSD and psilocybin suggest that by binding to serotonin 2A receptors psychedelics interfere with the PCC's ability to generate these low-frequency electrical oscillations. So rather than controlling the stream of top-down predictions, now anything goes. Brain activity becomes more entropic, reducing the influence of established networks and opening up new channels of communication.

Tellingly, psychedelics have two distinct, alternative effects on perception, depending on how noisy or unpredictable the sensory environment is. They either cause hallucinations or make ordinary perception more vivid. With eyes closed or in the pitch-black interior of a maloca during an ayahuasca ceremony, for example, people are likely to experience hallucinations. Under these circumstances the perceived trustworthiness of weak, noisy sensory data is upgraded, leading to odd, idiosyncratic inferences about what's happening.

My own experience shortly after drinking ayahuasca at the ceremony I attended near Iquitos in Peru seems to bear this out. After swallowing a modest dose of the bitter medicine, as night fell I distinctly heard the rhythmic sound of someone shaking a chakapa directly behind me, even though I knew the actual source of the sound was the shaman sitting hidden somewhere in the darkness in front of me. Her wistful song came from there, but I distinctly heard the rhythmic rattle of her chakapa coming from the opposite direction. Was the unchecked upwards flow of sensory information through my brain leading to false inferences about the faint echo of that dry, rattling sound after it bounced off the pillar I was leaning against?

In a brightly lit environment with eyes wide open, by contrast, psychedelics can enhance the vividness of sensory experience. In this case, strong, reliable visual data is given a surreal intensity. Aldous Huxley describes the phenomenon in *The Doors of Perception* – an account of his first mescaline trip in 1953 – as 'seeing what Adam had seen on the morning of his creation – the miracle, moment by moment, of naked existence'.[26] Three flowers in a glass vase on his desk were transformed, shining with brighter colours, meaning and significance. The same was even true of the books lining his study walls:

> Red books, like rubies; emerald books; books bound in white jade; books of agate, of aquamarine, of yellow topaz; lapis lazuli books whose colour was so intense, so intrinsically meaningful, that they seemed to be on the point of leaving the shelves to thrust themselves more insistently on my attention... I saw the books but was not at all concerned with their positions in space. What I noticed, what impressed itself upon my mind was the fact that all of them glowed with living light and that in some the glory was more manifest than in others.

Under mescaline's influence, with his eyes open in a brightly lit room, it was as if Huxley's brain was painting the books with intense colours and significance, and the books he loved the most shone more brightly than the rest. 'This is how one ought to see, how things really are,' he kept telling himself. He came to believe the drug 'cleansed the doors of perception', though a modern neuroscientist would be at pains to point out that on psychedelics, it's our prior expectations – built on the rich material of lifelong experiences and associations – that are embellishing the plain, unvarnished evidence of our senses. Hofmann wrote that after he had come down from LSD's powerful hallucinatory phase, in the afterglow it was as if the world was 'newly created'.

Huxley correctly intuited that:

> To make biological survival possible, Mind at Large has to be funnelled through the reducing valve of the brain and nervous system. What comes out at the other end is a measly trickle of the kind of consciousness which will help us to stay alive on the surface of this particular planet.

We can now be reasonably confident that in ordinary consciousness the 'reducing valve' staunching the exuberant flow of predictions from the 'Mind at Large' is the PCC, the key hub of the DMN. Unusual activity in this region of the brain and the wider network of regions of which it is a part has been implicated in a range of mental illnesses, including anxiety and depression. The network fires up whenever people get lost in the self-referential ruminations and worries – the stuff of autobiographical selfhood or ego gone awry – that characterize these conditions. Huxley had a neat turn of phrase to describe what it felt like when mescaline uncoupled conscious awareness from his ego: 'For the moment that interfering neurotic who in waking hours tries to run the show was blessedly out of the way.'

* * *

In the next chapter, I explore the idea that psychedelic drugs, by temporarily diminishing the network's influence, can facilitate healing in conjunction with talking therapies. When Albert Hofmann died in 2008 at the age of 102, the prospects for any further LSD research involving human subjects still looked grim. After a hiatus of half a century, however, the drug's fortunes began an extraordinary revival in 2016 with the publication of the Beckley/Imperial imaging study. A few days later, the overturning of a fifty-year ban on research was celebrated at Carlton House Terrace in London, home of the Royal Society, the world's oldest national scientific institution.

I was fortunate enough to be there that evening to hear the researchers talk about their findings. 'I think Albert would have been delighted to have had his "problem child" celebrated at the Royal Society,' Feilding told the audience of supporters, journalists and scientists. 'In his long lifetime, the academic establishment never recognized his great contribution to science and to humanity. But for the taboo surrounding LSD he would surely have won the Nobel Prize.'

She described meeting Hofmann in the nineties. 'Later I promised him that I would overcome the taboo and carry out scientific research with LSD in human subjects in time for his hundredth birthday in 2006.' Due to the considerable legal and psychological obstacles in her path, she said, she had missed her target by ten years, but finally the reputation of 'the jewel in the crown of psychoactive substances' was being restored. She called for LSD to be moved from Schedule I to Schedule II of the 1971 UN Convention on Psychotropic Substances to allow its potential therapeutic benefits to be explored further.

As I write, LSD remains in Schedule I alongside other potent psychoactive drugs judged to be a serious risk to public health and that have no recognized medicinal value. There are no signs this will change any time soon. Nonetheless, at least in the US, momentum is building for the rescheduling of psilocybin, a drug that Hofmann also played a pivotal role in bringing to scientific attention, isolating it from psilocybe mushrooms and publishing the results in 1958. A few months later his colleagues at Sandoz in Switzerland managed to synthesize the chemical for the first time. Six decades later in 2018, researchers at Johns Hopkins University – whose early results had suggested that it is highly effective at easing end-of-life anxiety – called on the US Food and Drug Administration (FDA) to reclassify psilocybin as a Schedule IV drug under US law, provided it passes muster in a larger clinical trial. This would put it in the same category as pharmaceuticals such as benzodiazepines and sleeping pills. The researchers point out that psilocybin is non-addictive and has no adverse physical effects.[27]

Even if all goes well, reclassification could take up to five years, but some US states may legalize medicinal use much sooner. In Oregon, for example, a measure to allow licensed growing of psilocybe mushrooms and psilocybin-assisted psychotherapy could be on the ballot in 2020 elections. Campaigns to decriminalize the consumption of magic mushrooms were boosted in 2016 when the annual Global Drug Survey found that they were the safest of all recreational drugs. As I write this in May 2019, citizens of Denver, Colorado, have just voted narrowly in favour of decriminalizing magic mushrooms, a move which will allow residents over twenty-one years of age to grow and consume them without fear of prosecution (buying and selling 'shrooms' will remain illegal under local, federal and state law).

I asked David Nutt, professor of neuropsychopharmacology at Imperial College London and a senior member of its psychedelic research group, how he imagined Hofmann would have felt about the belated recognition his work with psilocybin and LSD is receiving. 'He'd be honoured,' said Nutt. 'It's just a pity he's not here to witness his legacy. He saw the importance of LSD. He was the person who encouraged Sandoz to make it available to researchers around the world – he saw what a powerful tool it would be to explore brain function.'

No doubt Hofmann embarked on many more psychedelic adventures in the course of his lifetime, though presumably none would prove quite as frightening as his first, experimental LSD trip. His scientific and personal investigations led him to the belief that meditation and psychedelic drugs have similar beneficial effects on the mind and can work synergistically – a theme I explore in the final chapter. Three decades after he brought LSD into the world, Hofmann wrote in his book *LSD: My Problem Child*: 'I believe that if people would learn to use LSD's vision-inducing capability more wisely, under suitable conditions, in medical practice and in conjunction with meditation, then in the future this problem child could become a wonder child.'[28]

# 6

# *Mother Ayahuasca*

'I started drinking ayahuasca not for spiritual reasons but out of the desperation of being suicidal.' Benjamin Mudge is recalling the appalling trail of suffering that led to his trying the psychedelic brew for the first time twelve years ago at the age of thirty-five. Over the course of the preceding eight years doctors had prescribed no fewer than seventeen different pharmaceuticals – antipsychotics, antidepressants, mood stabilizers, anti-anxiety drugs – in a futile attempt to control the savage mood swings that characterize bipolar disorder. Things got so bad at one point Mudge was sectioned and spent time in a psychiatric hospital: 'I was on this horrendous wild goose chase of side effects trying to stabilize and get relief from extreme dysphoric mania alternating with extreme suicidal depression and total dysfunction.'

We are sitting at the dining table in the front room of his home in west London on a midsummer's day in 2018. About eighteen months before our meeting, Mudge relocated from Australia to the UK to care for his mother, a neuroscientist at University College London whose health had been deteriorating. As a young man, one of his ambitions had been to follow his parents, both professors, into academia, but bipolar put paid to those dreams. 'I was barely able to make it to the shops to buy a loaf of bread let alone be an academic.'

His condition was finally brought under control by the mood stabilizer lamotrigine, allowing him to hold down a full-time job for the first time in five years, but the side effects exacted a heavy price. 'My hair was falling out, I was twenty kilograms fatter than I am now and my emotional sensitivity was really warped in terms of my intimate communications with friends or a lover.' The drug seemed to rob him of any emotional awareness, he says. 'I was wearing a plastic smile.'

Once, before the stabilizers, antidepressants, antipsychotics and anxiolytics, music had been a passion, but the pills rendered him incapable of connecting emotionally with the instruments he once played with such enthusiasm. He could still play a little, but all the subtlety and soul were gone. 'If you listen to a guitar player who's technically quite proficient but there's no blues, no vibe, no soul, no jazz, it's kind of empty,' he tells me. 'That's how I was playing when I was on those pills.'

Music wasn't the only passion he lost: he used to be a painter. Mudge opens his laptop and shows me a painting of a man curled in a foetal position. The entire picture is suffused with an infernal crimson glow. It's labelled 'Self-portrait in a BAD mood'. 'I was a respected artist in my little community and that's what everyone expected I would do with my life. And then I took antipsychotics...' He pauses and corrects himself. 'Actually I didn't *take* them, I was *injected* with antipsychotics. I was locked in a straitjacket and literally injected in the butt with antipsychotics.' Ever since then he has been unable to paint. 'It's like the whole art department in my brain has shut down, or like forgetting how to speak a language. Really weird.'

Around 2 in every 100 people develop bipolar disorder. Their risk of suicide is between twenty and thirty times higher than average, with between a quarter and half of patients attempting to end their lives at least once.[1,2] And yet many refuse treatment. Mudge pulls up another slide from a presentation he gave in 2017 at the Breaking Convention psychedelic science conference in London, which was where I first heard his story. It's a compilation of photographs of famously creative bipolar patients who either took their own lives or whose deaths were related to their use of hard drugs or alcohol: Virginia Woolf, Kurt Cobain, Robin Williams, Amy Winehouse, Carrie Fisher, Heath Ledger, Van Gogh, Jimi Hendrix. 'What's currently available doesn't work for people like this whose creativity, spirituality, sexuality and emotional sensitivity are the most important things for them,' he says. 'These are precisely the things the pharmaceutical pills that are currently available numb or destroy.' He believes this is why creative people with bipolar often refuse treatment. 'It's a dilemma, because if you refuse treatment you get dangerously ill, then you start self-medicating.'

By taking the drugs prescribed him, Mudge was functioning reasonably well, and he still believes this to be the safest course of action for the majority of people with bipolar. But then life dealt him another cruel blow: in 2006 he was diagnosed with prostate cancer. Thankfully it was at an early stage. Many men will live with a non-aggressive prostate tumour into old age, often dying of another cause, but his doctor advised him that chronic stress was probably weakening his immune system and accelerating its growth. Rightly or wrongly, Mudge became convinced that by ingesting a daily cocktail of pharmaceuticals, enduring their physical and emotional side effects in order to hold down a job and do all the other things that were expected of him, he was hastening his own death.

'Against the wishes of my doctors and my family and most of my friends I went cold turkey,' he says. It's not something he would recommend to anyone. The withdrawal effects were shocking: not only self-harming and suicidal urges, but also aggressive impulses. 'And I'm a peace activist, I'm a really chilled kinda guy!' He laughs. 'So I just locked myself in my room.'

Thanks to the dedicated care of a few friends who stood by him in his decision to come off his meds, he weathered the withdrawal effects and began to cast around for alternative therapies for his condition. He consulted practitioners of Chinese medicine and tried more than a dozen different herbal remedies touted as treatments for depression, including St John's Wort and ginseng. None worked. 'Then someone said, "Have you heard of ayahuasca?" And I hadn't. They claimed it was a great antidepressant.' At that point Mudge had never dared try any kind of psychedelic. 'I was scared of magic mushrooms and LSD. Never touched them. But I was desperate, so I decided to try ayahuasca.' Mudge was living in the UK at a time when hardly anyone had heard of the almost mythical Amazonian brew, so it proved a huge challenge to track down anyone who drank it. 'The first time it helped a little,' he says, but the ceremony was not properly run so the whole experience was 'a bit of a disaster'. 'Then nine months later I discovered the Santo Daime.'

Santo Daime (pronounced *santo dye-me*) is a syncretic religion that uses the tea as its sacrament. Founded in Brazil in the thirties, outposts began springing up all over the world in the nineties. Even though it contains the banned psychedelic DMT, the US and Canada have legally sanctioned its use by bona fide churches such as Santo Daime and União do Vegetal. In several South American countries, its preparation, consumption and possession are legal, and not specifically prohibited in Spain, Italy, the Netherlands and

Australia. Santo Daime has churches in all these countries and several others where it operates beneath the radar of the authorities.[3] Crucially, the cooking technique and format of ceremonies are standardized across the world, providing a reasonably consistent brew and setting.

So it was in the reassuring environment of a Santo Daime church in London (since disbanded following a legal crackdown by UK authorities in 2012–13) that Mother Ayahuasca threw Mudge a lifeline. 'It worked really well for me so I started to drink ayahuasca there regularly. It could shift suicidality overnight and that could be sustained for at least a couple of weeks – feeling pretty good and more importantly feeling more like my true self, my authentic consciousness, my core personality.' With conventional antidepressant pills, any improvement in mood can take a while to kick in: a couple of weeks for MAOIs (monoamine oxidase inhibitors) and as long as a month for SSRIs (selective serotonin reuptake inhibitors). 'With ayahuasca I felt better within two hours. Significantly better.'

And so, like many other people with bipolar, Mudge is now 'self-medicating', though not with hard drugs or alcohol but with a tea prepared by macerating the stem of the vine *Banisteriopsis caapi* and boiling the resulting fibres with the leaves of the shrub *Psychotria viridis*. He says that by drinking ayahuasca every two or three weeks in a quiet, safe setting as part of a ceremonial ritual he has attained a 'humble happiness' – free not only from depression and suicidal urges, but also the out-of-control behaviour and grandiose fantasies that characterize mania.

After learning all he could from Santo Daime in London, Mudge moved back to Australia in 2007. For a while he lived contentedly in Byron Bay, New South Wales, a haven for alternative lifestyles where the jungle plants used to prepare the tea – imported years earlier by enterprising hippies – flourish in its semi-tropical climate. There was a Santo Daime community there and other ayahuasca drinkers who were brewing their own tea. 'I figured out this recipe that works really well for me, so I could have sat drinking ayahuasca on the beach on the dole for the rest of my life,' he jokes. But that wasn't his style and he felt a sense of responsibility towards all those others with bipolar in desperate need of an effective treatment free of emotion- and creativity-numbing side effects. 'All my favourite musicians kept committing suicide,' he says ruefully. In his former life he'd been a peace campaigner, founding the Peace Not War record label in 2001 and organizing fundraising concerts. Now he focuses all his energies on convincing the scientific and medical community that ayahuasca may be a viable treatment for bipolar, eventually winning a doctoral research scholarship in 2014 to conduct a proof-of-concept study at the psychiatry department of Flinders University in Adelaide.

When I first encountered Mudge, now forty-seven years old, I was intrigued but also sceptical. There is some preliminary evidence that ayahuasca is an effective treatment for major depression, but bipolar disorder? Everything I had read in research papers and in the media suggested that drinking the tea is risky for anyone with a family history of the condition, let alone someone who is themselves bipolar.[4,5] Psychedelics are widely considered to be 'psychotomimetic', temporarily evoking many of the symptoms of psychosis such as hallucinations and delusional thinking – symptoms that people with bipolar often experience during episodes of full-blown mania. Psychiatrists and most of the ayahuasca community therefore believe that for susceptible people to drink ayahuasca runs an unacceptable risk of triggering mania. Dozens of documented cases appear to bear this out. For the same reason, scientists recruiting volunteers for psychedelic studies bar anyone with a personal or family history of a psychotic illness from taking part.

Nonetheless I was eager to hear what Mudge had to say, having recently returned from my own experience of drinking a small dose of ayahuasca tea at a ceremony in the Dios Ayahuasca Sanaciones healing centre in Peru. Like Mudge I had taken a chance, albeit a smaller one: I have a family history of bipolar. I suffered no ill effects, apart from nausea, and enjoyed a beautiful, albeit unspectacular experience followed by a psychedelic afterglow of cheerful positivity that lasted several days.

But for someone diagnosed with bipolar to drink a psychedelic concoction with a reputation for triggering mania? That sounded more than a little crazy. Despite the weight of medical and scientific opinion, however, Mudge is convinced that in ayahuasca he has discovered the holy grail of bipolar treatment – a combination of drugs that prevents the debilitating mood swings between feverish, excoriating mania and suicidal depression while leaving one's creative and emotional sensibility intact.

Ayahuasca certainly seems to work for his brain, and Mudge thinks he knows why it sometimes causes mania in vulnerable people under particular conditions – conditions that in principle could be avoided if the treatment were ever to go mainstream. Supporting his contention is a small but growing number of anecdotal reports from other bipolar people who have benefited from drinking the tea. Among them is the British author Jay Griffiths, who has described travels with a friend in the Amazon during which she took part in several ayahuasca ceremonies. In *Wild: An Elemental Journey* she writes:

> The depression that had so darkened me for months had gone, and though during my months in Peru I had a

> persistent worry that it would return, it did not and I was free of it for years. I said my goodbyes to Jeremy and stayed on in the Amazon, my spirit as green, happy and elastic as a grasshopper in summer, tromboning in the grass.[6]

Other bipolar people Mudge has contacted speak of feeling confident, quietly happy and free of both mania and suicidal depression after drinking the tea. One case, who participated in five ceremonies, said that each had:

> … progressively altered the neurochemistry in my brain to such an extent that, with maintenance, I consider my bipolar illness to be in remission. My mood is more stable, I'm not having more depression or more mania at all, but there is an emotional process of integration going on in the days afterwards… The spiritual element has a huge part to play in terms of my emotional state – there is a sense that I am a living miracle, that life is a beautiful crazy ride, and an appreciation of that. But mania is a real ego-driven force that takes over the mind, and it's not like that at all. It's a general lightness and happiness, but not mania.

These are anecdotes, not data, and this bears repeating, but it is not unknown for medical revolutions to begin like this, with reports from patients desperate enough to try something new – anything to relieve their misery. Mudge has now initiated the long process of empirical investigation. There are two strands to his research. The first involves gathering reports, both positive and negative, about what happens when people with bipolar ignore the warnings and drink ayahuasca. The second follows a noble tradition in psychedelic investigation dating back to Albert Hofmann: Mudge is using himself as a guinea pig. His university's ethics committee has made it clear he can't give ayahuasca to anyone else, but it has no jurisdiction over what he does to his own body. Any results he gleans from self-experimentation will fall under the honourable title of 'autoethnographic research'.

'I go to an ayahuasca ceremony,' he explains, 'I drink the tea, have the experience, write down qualitative notes, and then I take home a teeny bit of the tea and put it through a freeze-dryer.' For each sample he knows the cooking time, which subspecies of *B. caapi* was used, the relative amounts of vine and leaf, and so on – factors that influence the ratio of the brew's four psychoactive ingredients. Apart from DMT from the leaves of *P. viridis*, ayahuasca contains three alkaloids from the woody stem of *B. caapi* vine: harmine and harmaline (both MAOIs), and tetrahydroharmine (an SSRI). Once Mudge has collected around fifty freeze-dried samples, he plans to have a lab determine the relative quantities of the four active constituents. He will then cross-check this data with his notes about each batch's effect on his mental state and attempt to pinpoint which constituent was problematic and which beneficial.

For now this is just a 'proof of concept' study, but eventually, when he has determined an optimum dosage and ratio of the active ingredients, he hopes that he or other psychedelic researchers will secure funding and ethical approval to conduct a Phase I clinical trial involving a small number of bipolar patients. This will determine the treatment's safety before larger trials can be conducted to test its efficacy.

The first stage of his research has progressed far enough for him to draw some tentative conclusions about why some bipolar people become manic after drinking ayahuasca, whereas others benefit. Among the fifty people whose reports he has documented, thirty said their symptoms improved, six had mixed experiences in different ceremonies (sometimes positive, sometimes negative), and fourteen saw a worsening of their condition. 'This is really what my PhD has boiled down to: how could it be that one substance has such remarkable effects in me and others, but in some bipolar people it triggers mania or even mania followed by depression?'

To understand what's happening, says Mudge, it's important to realize that the nervous systems of bipolar people are hypersensitive. Their mood is easily perturbed by environmental influences and psychoactive drugs, and this sensitivity is ratcheted up still further by the MAOIs in ayahuasca. These molecules inhibit the enzyme that would otherwise digest DMT in the gut, allowing it to pass intact into the bloodstream and onwards to the brain, but they are also antidepressants in their own right, ramping up neuronal signalling by slowing the breakdown of the neurotransmitters noradrenaline, dopamine and serotonin. The SSRI in ayahuasca (tetrahydroharmine) boosts serotonin levels. Excessive doses of MAOIs or SSRIs, common components of the drug cocktail that doctors prescribe for bipolar disorder, are known to cause mania. They also magnify the mind-altering effects of alcohol, tobacco, caffeine and recreational drugs.

So the hypersensitivity of bipolar people can lead to what Mudge calls 'false negatives' – adverse reactions that he believes could be avoided in a more controlled, therapeutic context. As with the blend of antidepressants, antipsychotics, mood stabilizers and anxiolytics doctors prescribe for bipolar, getting the dosages right and avoiding interactions with non-pharmaceutical drugs are key. 'Once you get rid of these false negatives it's actually incredibly good news,' he claims. For example, three of the patients who had negative or mixed experiences took part in multiple ceremonies over a short space of time, which would have led to a progressive build-up of MAOIs in their system. This need not present a problem for most people, but for someone with bipolar the resultant increase in neurotransmitter signalling could tip them over the edge into a manic episode. What's more, a tightly spaced series of evening ceremonies will probably have meant several sleepless nights in a row which, as we've seen, is in itself a risk factor for mania and particularly so for people with bipolar.

To prevent adverse drug interactions, many retreat centres strongly advise participants to observe a 'washout period', steering well clear of recreational drugs, alcohol and tobacco in the weeks leading up to a ceremony. This is especially important for anyone with a hypersensitive nervous system. But in ten of the twenty cases where bipolar people experienced negative or mixed reactions, Mudge has discovered they took other drugs before, during or after the ceremony, including psilocybin, mescaline, marijuana and *rapé* (pronounced *har-pay*) – a high-nicotine tobacco that is either smoked or inhaled as snuff in many indigenous ayahuasca rituals.

'If they're doing any of this long list of other contraindicated activities, *that* in my opinion is what has actually caused the majority of problem cases,' says Mudge. He also has a hunch that some of those who experienced mania unwittingly drank what he calls 'ayahuasca beer': a fizzy, fermented psychedelic brew cooked several months earlier that hadn't been refrigerated – perhaps because it had been sent abroad through the post or was carried around from ceremony to ceremony in a shaman's suitcase. As a result it contained alcohol. 'For most people that doesn't matter, but for bipolar people, that's the problem.'

These kinds of hazards are intrinsic to the way people currently drink the tea, often in ceremonies far from home, organized by strangers. Needless to say, Mudge's vision of the future of ayahuasca therapy for bipolar disorder doesn't involve patients waving their doctor goodbye and boarding a plane to South America every couple of months, or even attending their local Santo Daime church. Rather, they would drink a precise dose of a carefully prepared brew – containing the optimum balance of alkaloids and DMT – at a clinic with professionals on hand to help and counsel them. In the weeks leading up to the treatment they would need to get plenty of sleep and forswear antidepressants, recreational drugs, alcohol and tobacco. More controversially perhaps in the eyes of many psychotherapists, Mudge is convinced that rather than administering the treatment in a sterile, secular vacuum, there must also be ritual, spiritual imagery and symbols – perhaps reflecting patients' own belief systems – to help evoke meaningful, therapeutic insights.

There is good empirical evidence that spiritual experiences improve the chances of success in psychedelic therapy, a finding I explore in the next chapter. But does prescribing ayahuasca to treat bipolar make *biological* sense? It's a moot point. Some seventy years after it was first used to control mood swings, scientists still can't agree how lithium – the current mainstay of bipolar treatment – works. There are half a dozen theories about its mood-stabilizing effects (which incidentally only occur within a tight dosage range, above which lithium becomes toxic). What surely matters with any treatment is that it has been proven to work and that any side effects are tolerable. Nonetheless, there is a fascinating overlap between the molecular effects of lithium in the brain and those of DMT. One day this may help to reveal why all of us don't veer constantly between the emotional poles of elation and despair.

Neuroscientists believe the brain has evolved its own molecular mood-stabilizing mechanisms whose function is to dampen down emotional extremes. A good engineering analogy is a centrifugal governor, the ingenious device invented in the eighteenth century that prevents steam engines from running too fast or too slow, which was an indispensable component of the trains and factory machinery that drove the Industrial Revolution. Essentially the governor comprises two heavy metal balls attached by lever arms to a vertical spindle. As the engine speeds up, the spindle turns faster and the balls are displaced outwards until a predetermined point is reached when the mechanism closes a valve, reducing the flow of steam into the engine. As the engine slows, the rotating balls decelerate, the valve opens and the flow of steam increases once again. Like the multitude of homeostatic control mechanisms that maintain stability in our own bodies, this is a self-regulating, negative feedback loop.

The brain's mood-stabilizing governor will surely prove a lot more complicated, no doubt comprising a host of neurotransmitters, enzymes and receptors operating in concert, but when it works well it essentially performs the same function as the governor in a steam locomotive: keeping our emotional, cognitive running speed within safe limits. On the one hand it prevents our nervous systems from getting 'jammed' at top speed so we risk careering off the tracks, and on the other from grinding to a halt, unable to get moving again. These are the emotional extremes that make the lives of people with bipolar disorder so challenging.

Remarkably, Mudge's mother Anne may have made a crucial contribution to the search for better treatments for bipolar. Almost two decades ago, she and her colleagues at University College London demonstrated how lithium and other mood-stabilizing drugs affect a key component of this mood-stabilizing mechanism, known as the 'phosphoinositide cycle'.[7] Lithium and the other drugs act on a bewildering range of targets in the brain, so this was an important milestone. Anne Mudge had just finished working on a paper updating this research when she died, just a month before I interviewed her son – who helped get the article ready for submission to a leading neuroscience journal even as her health declined. It was published in August 2018 and shows that lithium and the antidepressant Prozac (fluoxetine) have opposite effects on the turnover of phosphoinositide in nerve cell membranes in the cortex: lithium slows it down and Prozac speeds it up.[8] This helps to explain why Prozac can trigger mania in bipolar patients unless they are also taking lithium. For these patients, taking Prozac is a bit like stepping on the brain's accelerator pedal, while lithium slams on the brakes.

Benjamin Mudge believes that a single dose of DMT during the depressive phase of his illness, in combination with the alkaloids in ayahuasca, is a gentler way to control the running speed of his brain. The advantages over lithium and antidepressants such as Prozac may be that DMT acts more quickly to ease depression, is more selective and doesn't have to be taken daily, preserving his creativity and emotional sensibility.

In common with the other classic psychedelics mescaline, psilocybin and LSD, DMT binds to and activates serotonin 2A receptors, which are responsible for these drugs' distinctive perceptual and cognitive effects. They are serotonin 2A 'agonists'. Crucially, however, DMT does not bind as tightly as the others, so its acute effects are more short-lived, which makes it less likely to trigger mania. This is why, as someone with bipolar, Mudge says he wouldn't touch psilocybin let alone LSD. But he has smoked pure DMT in his self-imposed role as research guinea pig without experiencing any ill effects.

If Mudge's hunch about DMT is correct, how does it stabilize mood? How does it lubricate the brain's centrifugal mood governor? For now the molecular mechanism remains a mystery, but there may be a clue in the astonishing range of mental afflictions that classic psychedelics are candidate treatments for. It's still early days, but clinical research is advancing rapidly in depression. At the time of writing, major trials of psilocybin as an antidepressant are getting under way. One is being conducted by the psychedelic research lab at Imperial College London. In 2017, the scientists published a pilot study in which they gave two doses of psilocybin one week apart to patients with treatment-resistant depression. There were remarkable improvements in symptoms that lasted at least three months and no serious adverse events.[9] But only twelve patients were treated and there was no control group so no way to determine whether the improvements were a result of the drug itself or simply a placebo effect. In theory, some of the patients might even have got better in the absence of treatment. In a more rigorous, 'randomized controlled trial', the Imperial team is now randomly assigning a much larger group of people with treatment-resistant depression to receive either a single dose of psilocybin or six weeks of a daily SSRI antidepressant.

Meanwhile, smaller studies hint at a plethora of other applications: LSD for the psychological distress associated with terminal illness; psilocybin for obsessive compulsive disorder (OCD), alcoholism and quitting smoking; and ayahuasca for unipolar depression, post-traumatic stress disorder (PTSD) and addiction.[10,11] Of course, Mudge would love to add bipolar disorder to the end of this long list. 'Panacea' is a word rarely justified in medicine, but if these treatments come to fruition it would suggest some kind of common denominator. Could psychedelics be fixing all these disparate psychological afflictions by greasing the mind's inbuilt mood governor?

In 1953, ten years after Hofmann first synthesized LSD, its transformative effects on consciousness and structural similarities to a recently discovered brain hormone called serotonin shook up scientists' ideas about how the brain works, revealing for the first time the power that neurochemicals hold over our perceptions, thoughts and emotions. Similarities between the hallucinogen and the hormone led directly to the idea that serotonin is involved in mental illness.[12,13] We now know that serotonin has a wide range of functions in the brain, affecting not just positive mood but also aggression, sleep, appetite, learning and memory. At the last count no fewer than fourteen different receptors for serotonin have been identified in the central nervous system, and two of them – dubbed '1A' and '2A' – may be crucial levers in the brain's mood governor.

According to a hypothesis recently proposed by Imperial scientists David Nutt and Robin Carhart-Harris, the mammalian brain has evolved a pair of counterbalanced strategies for dealing with challenging circumstances.[14] Levels of serotonin are known to rise under stressful conditions, increasing binding at all fourteen receptors, but Nutt and Carhart-Harris believe that 1A and 2A play a pivotal role in determining which strategy the brain deploys.

The first strategy, mediated by the more numerous 1A receptors, mitigates the effect of mild, everyday stress by damping down aggression, impulsivity, impatience and anxiety: it promotes tolerance or passive acceptance when things don't work out how we expected. This makes intuitive sense because when serotonin binds to 1A receptors on nerve cells it has an inhibitory effect, making them less likely to fire. To give an idea of the direction in which this pushes our mood governor, the drug ecstasy (MDMA) – which creates a relatively short-lived burst of serotonin in the brain – produces its 'loved-up' effects by ramping up 1A receptor activity. The long-term antidepressant effects of SSRIs are also thought to work through increased serotonin binding at 1A receptors.

I met Nutt in 2016 at his lab on the Hammersmith campus of Imperial College in west London. 'Low serotonin states are disturbed states,' he told me. 'People need serotonin: it immediately calms them, it immediately relaxes them. It gives them resilience.' In adulthood, the second stress-response strategy, mediated by the more sparse 2A receptors, only becomes apparent during the massive release of serotonin that occurs when we are in mortal danger – anything from the threat of being shot (the modern equivalent of being stalked by a predator) to starvation and asphyxiation. In contrast to 1A receptors, said Nutt, binding at 2A receptors is known to have a *stimulatory* effect on neurons. And in the long term, it promotes neurogenesis, the growth of new nerve cells.

This stress-response strategy of last resort has the effect of boosting sensitivity to environmental influences and facilitating relearning. It promotes behavioural and cognitive flexibility or 'plasticity', helping us ditch old established habits in favour of creative solutions that might just save our lives. Faced with almost certain death through starvation, for example, it might have allowed our ancestors to develop new hunting or foraging strategies. There is also evidence that increased serotonin 2A receptor signalling makes it easier to *unlearn* the dysfunctional fear conditioning that is at the root of phobias and PTSD (something which, as we saw in Chapter 2, REM sleep also does – see pages 42–43). These are exactly the kinds of therapeutic effect evoked by classic psychedelics, which of course bind strongly to 2A receptors. By simulating the deluge of serotonin that accompanies life-threatening stress, psychedelics appear to 'reboot' the brain, providing an opportunity to unlearn the maladaptive programming that underlies addiction, depression, PTSD and OCD, and replace it with healthier patterns of thought and behaviour.

If Nutt and Carhart-Harris are right, each of these mental afflictions is symptomatic of a failure of the brain's inbuilt mood governor, because despite the considerable stress and suffering these conditions have caused patients, relearning has not been triggered naturally. For whatever reason, these patients' governors are jammed. There is some evidence of this from post-mortem studies of the brains of unmedicated depressed patients and suicide victims, which contain unusually high numbers of serotonin 2A receptors. According to Nutt and Carhart-Harris, the brains of people suffering from depression may attempt to compensate for a lack of 2A signalling (and failure to trigger the behavioural, neural plasticity that might help) by churning out more 2A receptors. In other words, there has been 'adaptive upregulation' of the receptors in response to deficient 2A signalling. Their brains are effectively trying to initiate the second, more radical stress-response strategy. That the numbers of these receptors are still high in untreated depressed patients and people who have committed suicide implies that boosting the number of receptors hasn't worked, because a major increase in 2A signalling would in turn have 'downregulated' these receptors to more normal levels.

The idea that the relative influence of these two serotonin receptors helps determine our emotional and behavioural outlook in times of adversity – balancing the forces of conservatism and revolution – complements the entropic brain hypothesis proposed by Nutt, Carhart-Harris and their colleagues in 2014, which I introduced on page 127. It provides a molecular control mechanism for the heightened brain entropy (increased irregularity or unpredictability of connectivity between and within networks) observed in childhood, the psychedelic state and psychosis. According to the entropic brain hypothesis, healing during adulthood in conditions such as depression and addiction is made possible when order temporarily breaks down, providing an opportunity for undoing all that dysfunctional conditioning. Like a red-hot iron bar in a blacksmith's forge, in a state of high entropy our brains are more malleable: more open to change.

Nutt's taste for straight-talking about drugs and their legal classification endeared him to the British press in his days as the government's chief drugs adviser – a post he was sacked from in 2009 after suggesting ecstasy was less dangerous than horse riding, and LSD less of a threat to public health than tobacco or alcohol.[15,16] When I interviewed Nutt he was in a characteristically relaxed, expansive mood. 'We think the serotonin 2A receptor is critical for increasing brain plasticity,' he explained, leaning back in his chair and putting his feet up on the desk. 'In times of extreme stress or the provocation of psychedelics it can break down traditional patterns of thinking.' One line of evidence for this, he said, is that 2A signalling plays a critical role during infancy, a period of intensive learning and exceptional openness to experience. 'The young brain might actually be resistant to psychosis,' he mused. 'We don't know. Babies in that state of wonder are probably having a kind of psychedelic experience all the time.'

In adults, such wide-eyed levels of 2A signalling only occur in life-threatening adversity, in psychosis or on psychedelics:

> In extreme states, maybe in childhood and maybe when you're very stressed as an adult, only in those states will you get enough serotonin released to turn on those receptors. But when you do, your brain changes. It may change for the good or it may change for the bad.

This helps to explain why maladaptive thoughts and behaviours can get etched into the young brain during an abusive childhood or at times of extreme stress in adulthood, such as an assault or in combat. These experiences can lead to anxiety and depression in later life or even PTSD – ironically the very conditions that psychedelic-induced 2A signalling and plasticity may later help people to *un*learn.

According to Nutt and Carhart-Harris, psychedelics' ability to increase plasticity is what makes them promising treatments for such a wide range of mental illnesses, though it's important to distinguish between the effects people experience while under the influence of the drugs, and chronic effects in the ensuing days and weeks: the 'afterglow' of positivity, friendliness and openness that many people report experiencing. The scientists believe the most important acute effect is increased neural, cognitive plasticity (via a temporary burst of serotonin 2A signalling) – providing a therapeutic opportunity for radical change – whereas the longer-term benefits are improvements in mood and more creative responses to life's challenges. So while conventional antidepressants have a reputation for dulling creativity and blunting emotional responses to stress (through increased serotonin 1A signalling), psychedelics may have the opposite effect.

A healthy brain needs to retain some capacity – perhaps by fine-tuning serotonin neurotransmission – to adjust the balance between order and flexibility to meet changing circumstances. Is this the mood governor that malfunctions in people with bipolar disorder, causing precipitous swings between the low-entropy, inflexible state that is depression, and the high-entropy, plastic state that is mania? Like psychedelics, lithium is known to restore neural plasticity, which is compromised in people with bipolar disorder, and there is some evidence that both lithium and DMT stabilize serotonin transmission. Research remains at an early stage – much of it of necessity conducted on animals rather than humans – and the results to date have been equivocal. However, both molecules do appear to affect the balance of serotonin 1A and 2A receptors in the brain. Long-term lithium administration has been found to raise serotonin levels in the hippocampus of rats, which in turn downregulates their 2A receptors. Similarly in humans, in the happy afterglow following a psychedelic trip that can last for several weeks, 2A receptors are also known to be downregulated.[17,18]

If ayahuasca were ever to become a mainstream treatment for bipolar, one of the major advantages over conventional mood stabilizers would be that patients only need drink the tea intermittently, at most every couple of weeks. Drugs such as lithium and lamotrigine have to be taken daily, year in and year out. So while a dose of DMT may lubricate the brain's centrifugal mood governor then leave it alone, allowing a person's emotional 'running speed' to range upwards and downwards within normal limits, the daily drip, drip, drip of a pharmaceutical mood stabilizer may gum it up, restricting its scope to a very narrow emotional range. Whereas conventional treatments numb emotional responses in people with bipolar, DMT may preserve them. This would explain Jay Griffiths' joyous description of feeling 'green, happy and elastic as a grasshopper in summer, tromboning in the grass' for months after drinking ayahuasca. It might also help explain Benjamin Mudge's observation that, whereas the cocktail of drugs his doctors prescribed seemed to rob him of his musical, artistic mojo, ayahuasca is steadily restoring it, leading him back to his 'true self'.

Mudge knows he has a long way to go in his quest to convince psychiatrists that ayahuasca could make a safe and effective treatment for bipolar disorder. But if he succeeds in changing minds and winning funding, perhaps in a few years there'll be a clinical trial. In the meantime, he finds it profoundly distressing whenever a suicidal fellow patient reaches out to him for help and there's nothing he can do. To stay within the strict limits demanded by research ethics, he must continue experimenting on his mind and his mind alone. He told me:

> I am using myself as guinea pig, very consciously, very deliberately, in service to other people. I have survived extremes of suicidal tendencies and I know how to get through these things. I know when they're coming on and I know how to reach out for help. I've been manic quite a few times and, basically, I can handle it. I'm making the choice to risk a manic-depressive mood swing for the sake of discovering an ideal medicine that could help thousands of other people like me who are treatment-resistant, suicidal – stuck.

Just a decade ago his championing of such a radical alternative therapy would have seemed at best naive and at worst dangerous. But in the past few years we have witnessed a sea change in attitudes, with psychoactive drugs formerly demonized by doctors, politicians and the media now being recognized as treatments that can potentially alleviate human suffering where conventional treatments are failing. In the US, to date all but three states have legalized medical cannabis since California led the way in 1996. The street drug ecstasy is also undergoing a remarkable rehabilitation. In 2017, after impressive results in early clinical trials, the US Food and Drug Administration (FDA) designated MDMA-assisted psychotherapy for severe PTSD a 'breakthrough therapy', meaning it will be fast-tracked through Phase III trials so that, if successful, the treatment can be made available to patients as soon as possible. Worldwide, several clinical trials of psilocybin-assisted therapy for anxiety and depression are under way, and in October 2018 the FDA granted it breakthrough therapy status for treatment-resistant depression. The following month, the UK government moved medicinal cannabis products from Schedule 1 to Schedule 2 of the Misuse of Drugs Regulations in recognition of their therapeutic value in conditions such as multiple sclerosis and intractable epilepsy.

In light of this renaissance and his own, admittedly preliminary research, Mudge's radical proposal deserves to be taken seriously.

# 7

# *Death of the Ego*

Music orchestrates the experience, guiding us from emotion to emotion, from major to minor, from ecstatic highs to crushing lows and back again in the unravelling story of our selves.

It begins with ambient electronic music, melancholic and magisterial, establishing a calm, reflective mindset. Around forty minutes after consuming the truffles, the experience has changed, though I couldn't say exactly when or how. Coloured points of light are surging and swirling through a transparent cube revolving slowly in the blackness behind my blindfold. To my disordered mind, the lights and the cube are driven by the scientists' musical, brain-scouring healing machine. Collaborating somehow in its sentient mechanism, the smell of incense, pungent and oppressive, floods my senses.

Around fifty minutes into the experience and the rhythm and urgency of the musical light show accelerate. I realize this is it: lift-off. I do my best to surrender to the machine, but dread surges inside me.

A silence follows, like dreamless sleep, about which I remember nothing. When consciousness reawakens with the music, inside a bright void double basses induce the grinding despair of annihilation. Its funeral soundtrack is the opening of Henryk Górecki's

'Symphony No. 3'. Yet there is no fear in this place, only mourning, and steadily the musical mood is lightening, transcending the terrible thing that has happened.

In the bright void the question somehow arises: *Who is James?* The name summons him and, as if from the outside, *whatever remains* senses the entity that was James, his entire life encapsulated in a single impression, just as those close to him must have experienced him. At first, *whatever remains* mourns him with loud, uncontrollable sobs of grief, but later, as the music lightens further, come tears of relief. He wasn't so bad.

Above all, this place beyond life and death feels utterly real, and in the bright void a word evokes everything. Still outside my self but thinking clearly, *whatever remains* now contemplates all the vibrant, vivid entities ever personified by the name 'Kingsland'. I watch and reflect on them as they blink in and out of existence. Like me they leave behind nothing but the faint trace of memory.

A lifetime or perhaps just a few minutes have passed when a lull in the symphony's first movement allows me to resurface into external reality, just moments before the soprano begins her familiar lament of loss and mortality. I have already crossed this dreadful threshold and I don't want to be dragged back. To escape, I sit up and pull off my eye mask, find my glasses and put them on. I start rising to my feet to go out into the sunlight, but a gentle hand on my shoulder restrains me and so I must allow the song to double me up in mourning all over again, this time physically. And as my head sinks between my knees it feels as though a lifetime of tension is draining away until nothing is left.

At long last the soprano's mournful song is over. I straighten up, open my eyes and look in astonishment at the guardian angel kneeling by my side. She asks me if I'm OK and I say, 'All is well', and for once believe it. 'The Shepherd's Song' from Canteloube's heavenly *Songs of the Auvergne* is now playing inside the ceremonial hall and through a speaker outside on the grass in the sunlight. A little unsteadily, I get to my feet and walk out into the garden which, to my amazement, while I was dreaming of death has been magically transformed into Eden.

In everyday consciousness, music can evoke just the right mood for the occasion – for a party, a wedding, a funeral, a TV commercial. Under the influence of a psychedelic drug, this capacity of music to induce a particular emotion, especially if the music is to our taste, becomes yet more powerful. The synergy between drug and music has been exploited to brilliant effect by Mendel Kaelen of Imperial College London to create a therapeutic journey for people suffering from intractable major depression. His playlist guides patients through a calming landscape of plains and foothills interrupted by intense, emotional peaks that arouse the overpowering, often painful autobiographical memories and visual imagery that are the key to effective psychotherapy.

I was fortunate to have Kaelen's soundtrack playing as I and a dozen new friends ingested psilocybin-containing magic truffles on a 'psychedelic experience weekend' in the early summer of 2018. The setting for this experience, expertly organized and supervised by the UK's Psychedelic Society, was a cluster of single-storey buildings about an hour's drive from Amsterdam set in several acres of lovely gardens – a huge lawn, flower beds and a pond seething with tadpoles surrounded by acres of woodland.

Magic truffles are legal in the Netherlands. You can buy a 15g vacuum pack of the sinewy nodules of mycelium for around €25 from one of the many 'head shops' in the tourist district, as I did one afternoon about eighteen months earlier. As a speaker at the evocatively named Brainwash Festival, I'd just given a talk about mindfulness to around 200 earnest young people in a church by a canal. That evening in my hotel room, after laboriously chewing and swallowing around half of the dry, rubbery-tasting fungus (imagine eating a ground-up flip-flop), I began listening through headphones to my own calming playlist under the watchful eye of my partner. A couple of hours later, after an enjoyable but underwhelming experience, I munched and swallowed the remaining contents of the pack, which was how I came to be floating disembodied among the rafters of a lofty, cathedral-like space, conjured by my temporal lobes, my visual cortex and John Rutter's celestial 'Veni Sancte Spiritus' sung by the Choir of King's College, Cambridge with organ accompaniment.

Unlike Kaelen's, my playlist was emotionally unchallenging. It was a stroll over gentle rolling hills. There were no dangerous, craggy peaks to scale. This goes a long way towards explaining why my later trip, guided by Kaelen's healing soundtrack, was so much more interesting. That and the fact that we consumed 30g of truffles, twice the previous dose. Nausea and occasionally vomiting can be a problem shortly after eating psilocybin truffles, so most of my fellow psychonauts were advised to ingest them as an infusion, with a side order of ginger tea. But I hadn't been particularly bothered by nausea on my earlier trip, and when I described it as 'tame' to Stefana, one of the Psychedelic Society's three expert facilitators (our guardian angels), she recommended chewing and swallowing the truffles rather than drinking them as a tea. She explained that this would give me a slower climb to a more intense peak, followed by a gentler descent.

The result was the harrowing but nonetheless therapeutic trip described in the opening paragraphs of this chapter, without doubt one of the most insightful experiences of my entire life. There was death, but also redemption. For a few minutes this fungal molecule had wrenched me out of myself and exposed the world and my little place in it, revealing everything in its true perspective. It was brutal, but by the time I walked out into the garden the word 'FORGIVEN' had impressed itself so strongly on my consciousness it was as if the letters were written large across the sky. I had no idea I'd been carrying around so much guilt. To finally let it all go was an indescribable joy and relief.

I should emphasize that I have reconstructed what happened – the approximate timings, what music was playing and when – by listening again to Kaelen's playlist in a more sober state. While my consciousness was in thrall to psilocybin, I had no sense of the passing of time or any interest in the labels we normally use to define music, such as its title, genre or even what instruments are playing. And anyway, as several of my fellow psychonauts put it later, we couldn't tell whether our mental imagery was being generated by the music, or our minds were generating the music. We were completely, synaesthetically absorbed in it all. The music and our visual, autobiographical, emotional experience were indistinguishable.

In the years after Albert Hofmann introduced LSD to the world, psychiatrists were quick to recognize the potential of psychedelic drugs as an adjunct to therapy for conditions such as alcoholism, with music being used from the outset in the belief it not only provided a reassuring framework for the experience but also enhanced patients' emotional responses, mental imagery and recall of personal memories. Just a few years ago, as a PhD candidate at Imperial, Kaelen became the first scientist to rigorously test these assumptions about the role of music in psychedelic therapy.

He confirmed that LSD enhances emotional responses to music, and later discovered that among patients given psilocybin for treatment-resistant depression, improvements in their symptoms one week later correlated with how open they had been to the music and how well it had resonated with them, rather than the intensity of the drug experience per se. It was the interaction between the drug's effects and the subjective influence of the music that really mattered.[1,2]

In other studies, Kaelen and his Imperial colleagues have confirmed that music increases the intensity of complex mental imagery that people on LSD see with their eyes closed. They have also used functional magnetic resonance imaging (fMRI) to reveal how this alliance of music and drug alters brain connectivity during the trip. Music enhances volunteers' visions of objects, people and scenes, and these visions are associated with an increased connectivity and flow of information between a region involved in encoding and recalling memories – the parahippocampus – and the visual cortex.[3] The parahippocampus is part of the temporal cortex – the core of the brain's prediction processing hierarchies – and a component of the default mode network (DMN), which helps create our sense of selfhood. So it may serve as a 'connector hub' uniting memory, vision and selfhood. It makes intuitive sense, therefore, that abnormal activity in the parahippocampus has been implicated in psychosis, depersonalization and the 'dreamy' state of temporal lobe epilepsy.

Kaelen described his work, and the synergies between drug and music it has revealed, to an audience at a Royal Society event in London I attended in April 2017 celebrating the overturn of a fifty-year ban on human LSD research. 'Only when music was present was there an increase in information flow between the parahippocampus and the visual cortex,' he explained. This brain circuitry is known to be responsible for playing back autobiographical memories with accompanying mental imagery. 'The interaction between LSD and music seems to hijack the system and increase information flow.'

Psychedelics set in motion their dramatic effects by binding to serotonin 2A receptors in cortical regions high in the brain's processing hierarchies, in particular within areas comprising the DMN. This opens the floodgates – the neural barriers that normally limit the downwards flow of sensory, cognitive predictions from the temporal lobes – unleashing a turbulent stream of visual, emotion-laden memories: the precious material that psychiatrists believe is crucial for healing in addiction and other mental illnesses. In common with other altered states of consciousness, such as dreaming and hypnosis, the relearning of deep-rooted, maladaptive ways of thinking and behaving takes place when the usual sensory and cognitive restraints have been removed. Only by temporarily making the mind more *suggestible* can the accumulated detritus of a lifetime of experience be flushed away.[4]

As we've seen, psychedelics increase the entropy or unpredictability of brain activity, reducing the internal integrity of its networks and promoting cross-talk between regions that don't usually communicate. The drugs work like champagne at a party, breaking up the knots of guests who already know each other and increasing the likelihood of conversations between former strangers. What Kaelen's work has revealed is that playing some good music at this cerebral party increases chatter between the parahippocampus and visual cortex.

Music is also known to create the right mood for changes in personality on psychedelics. Alexander Lebedev, a psychiatrist at the Karolinska Institute in Sweden who collaborated with the Imperial psychedelics lab, compared MRI scans of volunteers who were played music while under the influence of LSD and those who were not. He found that, overall, acute increases in entropy in their brains were associated with increases in the personality trait 'openness' two weeks later. This dimension of adult personality is associated with characteristics such as imagination, creativity, aesthetic appreciation, non-conformity and willingness to try out new experiences. But when music was playing, increases in entropy were even more strongly associated with subsequent improvements in openness. Importantly, as will become clearer later, the clearest link between entropy and personality change was found among those who experienced the greatest degree of ego dissolution while listening to music.[5]

When I met David Nutt, professor of neuropsychopharmacology at Imperial College London, he explained the possible long-term therapeutic benefits of increased openness, which in the natural course of events doesn't change very much once a person reaches adulthood. 'It may be that if you're more "open" you're more open to learn,' he said. 'You will accrue new thoughts, new experiences and new insights because your brain isn't locked into the traditional ways of thinking.'

Psychedelic therapy offers people a window of opportunity to change their lives that wouldn't otherwise be available to them. Music plays a vital role in throwing open this window, said Nutt, provided it resonates with their personal tastes. Otherwise it doesn't work so well. One patient in the Beckley/Imperial pilot study investigating psilocybin as a treatment for depression, for example, had a strong dislike for the modern music on Kaelen's playlist, which provoked frustration, irritation and resistance. My feeling of dread listening to some of the synthesized music early in the playlist may be another instance of this kind of idiosyncrasy. Rather than providing calm and reassurance, parts of the electronic soundscape aroused in me the sort of futuristic unease evoked by the soundtrack of the sci-fi movie *Blade Runner*, an overpowering sense of helplessness in the face of an alien, machine intelligence – though as far as I know none of my fellow psychonauts felt the same way. 'One of the questions [with psychedelic therapy for depression] is whether you should personalize the music,' said Nutt. 'We haven't yet, but we could certainly do that.'

The kind of music we love *speaks* to us, summoning the emotional connotations we have learned to associate with it over the years. Neuroscientists have found an extensive overlap between the cortical regions that the brain uses to predict the meaning of spoken words (and sign language) and those it uses to interpret the emotional meanings of music. This overlap may stem from the way emotion was first communicated to us as babies through the timbre of our mother's voice. Psychedelic molecules bind to the profuse serotonin 2A receptors in these parts of the cortex, in particular the insula, where awareness of all our feelings and emotions seems to arise, and in the planum temporale, which is adjacent to the auditory cortex and a key component of the brain's language-processing circuitry. This helps explain how drugs like LSD and psilocybin amplify the very personal emotional messages already pinned to particular kinds of music. Kaelen writes in his PhD thesis that psychedelics may 'tune' the brain to the acoustic properties of music that carry this emotionally charged information.[6]

The overlap between the neural signatures of music and language – they activate the same parts of the brain – strengthens the case made by some anthropologists that, hundreds of thousands of years ago, music preceded the development of language as a channel of communication helping to bind increasingly large groups of prehistoric humans together on the African savannah.[7–10] If so, it seems likely that music played a central role from the outset in religious and healing rituals during which consciousness-altering substances were consumed. Chanting, songs, rhythmic clapping, drumming and percussion remain a universal feature of indigenous psychedelic ceremonies all over the world, such as the Native American shamanic ceremonies that use mescaline-containing cacti; the ayahuasca ceremonies of South America; and the psilocybe mushroom ceremonies of Central America, Siberia and Papua New Guinea. Like the playlists tailored for modern psychedelic therapy, music in these traditional settings focuses the attention of participants, provides a safe framework for the experience and evokes healing emotions and visions. Shamans in South America, for example, have said the *icaros* they sing and whistle during personalized ayahuasca ceremonies are designed to ease anxiety and summon the visions that they believe will resolve a particular client's problems.[11]

Kaelen recommends music that sounds unfamiliar or exotic, ideally without words or in a language unknown to the patient, because attending to the meaning of lyrics can be a distraction: an anchor preventing the mind from setting sail for deeper waters. According to the veteran psychedelic researcher Roland Griffiths, the right kind of music can help trigger the mystical experiences that are so important for healing in conditions such as addiction and end-of-life anxiety. He and his team at Johns Hopkins University in Baltimore, US, have identified the key acoustic features of pieces favoured by psychedelic therapists around the world to evoke a 'peak experience' in which ego boundaries break down. Among the tracks the therapists recommend are 'In a Silent Way' by Miles Davis, 'Echoes' by Pink Floyd, Samuel Barber's 'Adagio for Strings' – and Górecki's 'Symphony No. 3'. Griffiths and his colleagues write that their analysis suggests an ideal track for summoning mystical experiences has a slow tempo, 'regular, predictable, formulaic phrase structure and orchestration, a feeling of continuous movement and forward motion that slowly builds over time'. They advise against jarring transitions or a lack of predictability, which can provoke feelings of unease.[12]

You may be forgiven for thinking that all this sounds a little clinical. For anyone who has not taken a high dose of a psychedelic, it is easy to dismiss the mystical experiences people report having on these drugs – perhaps while listening to carefully selected music – as less insightful or 'real' than the spiritual epiphanies that strike suddenly out of the blue as a result of more conventional, mainstream religious practices such as prayer, contemplation and fasting. They might also doubt, perfectly reasonably, whether drug-induced insights could precipitate enduring changes in a person's outlook and life choices.

These questions were first tackled six decades ago by a student of the history and philosophy of religion at Harvard University called Walter Pahnke in a research project for his doctoral thesis.[13] Pahnke, who was also a Christian minister and a physician, famously gave psilocybin or a placebo to theology students shortly before the start of a Good Friday service at Boston University's Marsh Chapel in 1962. The objective was to find out whether the peak experience of someone who has ingested a classic hallucinogen as part of a familiar religious ceremony is any different from experiences described in the mystical traditions of Christianity, Islam and Judaism.

The latter are often packaged in the poetic language of another age. Writing in the sixteenth century, the Spanish Carmelite nun St Teresa of Ávila described the fleeting glimpse of God's glory afforded to someone practising the disciplines of solitude, mortification and prayer as 'a precious pearl' and 'greater than all the joys of earth'. She wrote that experiencing the union of one's soul with God was 'a delectable death' and compared the transformation that followed to a silkworm changing into a butterfly.[14] In a similar vein, the fourteenth-century anchorite Julian of Norwich wrote that, 'Prayer uniteth the soul to God. When He of His special grace will shew Himself, He strengtheneth the creature above its self…' Perhaps the best known of the sixteen divine revelations that came to Julian as a young woman while she was feverish and close to death was that, 'All manner of things shall be well… the least thing shall not be forgotten.' In another revelation, she was shown a round object in the palm of her hand the size of a hazelnut which, she was told, contained 'all that is made'. It had three properties: 'God made it… God loveth it… God keepeth it.'[15]

How does one bridge the cultural and religious divide between the spiritual experiences of Christian mystics like St Teresa and Julian of Norwich, and those of, say, a Sufi mystic? Pahnke began by drawing up a list of nine fundamental characteristics of mystical states of consciousness that most religious scholars agree to be universal. The characteristics were:

1. A sense of unity (either internally, or with people, supernatural beings or objects in the outer world).
2. Transcendence of time and space.
3. Deeply felt positive mood.
4. Sacredness.
5. Insight into 'ultimate reality'.
6. Paradoxicality (the experience is logically contradictory, for example there is a loss of selfhood and yet an observing self remains).
7. Ineffability (beyond words).
8. Transience.
9. Persisting positive changes in the individual's attitudes and/or behaviour.

He then recruited twenty male, Christian seminary students with no prior experience of taking psychedelics. After several prep sessions in the preceding weeks under the guidance of group leaders

– designed to foster trust and reduce anxiety – the experiment itself took place in a prayer chapel in the basement of Marsh Chapel while the Good Friday service was held in the sanctuary above and broadcast to the smaller chapel through loudspeakers. Ninety minutes before the service began, in a side room, each of the volunteers was handed a blank envelope containing a capsule. Half of the capsules contained 30 milligrams of psilocybin and half contained 200 milligrams of nicotinic acid, a vitamin that isn't psychoactive but causes a temporary warmth and tingling of the skin. At this stage none of the research assistants supervising the experiment or the students themselves knew who would take the drug and who would take the placebo.

Shortly before the service a bell tolled summoning the theology students to the dimly lit chapel, where they sat quietly listening to a prelude being played on an organ in the sanctuary above. The set-up in the little chapel would have been perfectly familiar to them: on the altar was a pair of lit candles with a gold crucifix between them, and behind the altar were two stained-glass windows. When the organ fell silent, the Reverend Howard Thurman, Harvard's charismatic chaplain, welcomed the students before going upstairs to start the traditional Good Friday service, a sombre fixture in the Christian calendar during which worshippers contemplate Jesus's sacrifice on the cross.

Pahnke wrote in his thesis that the service comprised singing, more organ music, Bible readings, personal meditation and prayers. He reported that little of note happened in the small chapel during the first half of the service, but in the second half there were tears and spontaneous exclamations from some in the congregation. Against all the usual conventions, a couple of students went up to the altar and one even entered the pulpit, sat at the organ and mimed playing. A few stretched themselves out on the pews, others on the floor. At this stage it wasn't hard to guess who had received the psilocybin and who the placebo, and it became even more obvious after the service. Those who had taken the drug had dilated pupils, wrote Pahnke.

> Experimentals were also more informally attired; ties had been loosened or removed, and hair was not combed. Their dominant mood was quiet detachment or joyful exuberance. They wanted to share the impact of their experience with their friends. Two had a little difficulty in readjusting to the 'ordinary' world and needed special reassurance by their leaders until the drug effects subsided.

Immediately after the service, participants were tape-recorded describing what had just happened, first in individual interviews and later in group discussions. They also provided written accounts and during the following week completed a questionnaire designed to elicit whether or not they experienced any of the nine universal characteristics of a mystical state of consciousness. Six months later each subject filled out a follow-up questionnaire and was interviewed once again and his responses recorded. The written accounts and transcripts of the recordings were then rated by independent judges (who were naive to the nature and purpose of the experiment) according to how well they matched the nine criteria for a mystical state of consciousness.

The description written by one particular student, who was among those who took psilocybin, serves to give a flavour of their responses:

> I cannot describe the sense of the Divine – He was the eternal mystery that was: He was everywhere, but completely

> transcendent; the Divine, truly not of this world, but whose message had the greatest significance for this world. I felt compelled to go to the front of the chapel, to minister in the name of Christ – for no one else was doing it, and it had to be done…
>
> Then I read the Scripture, put out the candles (which I believe to be symbolic of the crucifixion of Christ), and after more blackness, found myself in the pulpit, preaching about love and peace… I attempted to play the organ, wanting to play 'Christ the Lord is Risen Today', being motivated by a strange sense of joy in the reality of this event…
>
> Of first significance was the feeling of a profound religious 'call' – the first I think I have ever really had. Before, I just felt as if I should enter the ministry, but now I 'know' that I must… up to this time, I have never known the real meaning of the Christian truth – I have overintellectualized it, and have not involved myself in its eternal meaning and significance.

Having analyzed all the results, Pahnke's conclusion was emphatic. 'Under the conditions described,' he wrote, 'psilocybin can induce states of consciousness which are apparently indistinguishable from, if not identical with, those experienced by the mystics.' The controls, he noted, had met none of the nine criteria. He further concluded that, six months on, the experiment appeared to have had a profound impact on the lives of eight out of the ten subjects given psilocybin. 'These subjects felt that this experience had motivated them to appreciate more deeply the meaning of their lives, to gain more depth and authenticity in ordinary living, and to rethink their philosophies of life and values.'

Remarkably, some of these positive effects were still evident a quarter of a century later. A student called Rick Doblin, as research for his psychology undergraduate thesis at New College of Florida, tracked down nineteen of the twenty participants, sixteen of whom agreed to be interviewed and repeat the original six-month follow-up questionnaire.[16] While many persisting positive changes were reported by the experimental group, none was reported by the control group. Doblin – who would go on to found and direct the influential Multidisciplinary Association for Psychedelic Studies (MAPS) – wrote in 1991 that for those who had taken psilocybin 'the experience helped them to resolve career decisions, recognize the arbitrariness of ego boundaries, increase their depth of faith, increase their appreciation of eternal life, deepen their sense of the meaning of Christ, and heighten their sense of joy and beauty'.

That a single dose of a substance, administered to believers in a setting appropriate to their faith, could have such a profound, positive and lasting influence on their spiritual lives is astonishing, though as far as I'm aware magic mushrooms have yet to be considered by any mainstream Christian denomination as a third sacrament, alongside the bread and the wine. From a purely scientific viewpoint, however, there are good reasons to be sceptical of what became known as 'the miracle of Marsh Chapel'. Its claim to be a 'double-blind' trial (in which neither subjects nor experimenters knew who received the active treatment and who got the placebo) was incorrect because it quickly became obvious to everyone which seminary students were tripping and which were not. The subjective effects of nicotinic acid are completely different from those of psilocybin, their onset is much quicker and they don't last nearly as long. As a result, there would have been a much stronger expectation of an effect in the psilocybin group, potentially biasing the results. In addition, the independence of each student's report of the experience was almost certainly compromised because participants took part in the experiment as a group and were free to communicate during the service and afterwards.

In hindsight, perhaps the most serious flaw in the methodology of the Good Friday experiment was that half of the researchers who helped conduct it took psilocybin themselves that day. This wasn't Pahnke's fault, it was simply part of the culture of psychedelic research at Harvard – his PhD supervisor was none other than Timothy Leary. During Doblin's follow-up study, he also discovered that, perhaps as a result of the climate of hostility already closing in on psychedelic research at Harvard in the early sixties, Pahnke had failed to mention in his report that there was a scuffle when one of the theology students who took psilocybin tried to leave the building during the service and was restrained by the group leaders. Pahnke injected him with the tranquillizer Thorazine and he sat quietly for the rest of the service and appeared to suffer no long-term ill effects.

Due to the legal clampdown on psychedelic research that began later in the sixties, forty years would pass before more rigorous repeats of the trial could be undertaken. But when they finally were, the results largely supported Pahnke's startling conclusions. In 2005, for example, Griffiths and his colleagues published the findings of a double-blind trial that studied participants individually rather than as a group, and used methylphenidate (better known as Ritalin) as the control. This powerful stimulant has some similar psychological effects to psilocybin that kick in at roughly the same time after ingestion and last about as long, making it a much better control. During psilocybin sessions, 61 per cent of participants experienced a 'complete mystical experience' (they scored above a predefined threshold on a questionnaire similar to that used by Pahnke assessing factors such as feelings of unity, positive mood, ineffability and sacredness) whereas only 11 per cent did after taking methylphenidate. Two months after taking psilocybin, 71 per cent reported that it had been one of the five most spiritually significant events of their life, and 33 per cent said it was the most significant of all. By contrast, two months after taking methylphenidate, 8 per cent rated it in their top five for spiritual significance and none rated it as the most significant.

In a follow-up study, Griffiths showed that the likelihood of a complete mystical experience was greater the higher the dose of psilocybin, strengthening still further evidence that it is the drug that causes these experiences, and not simply the power of expectation or wishful thinking. Furthermore, the effects had an enduring influence on participants' mood, attitudes and behaviour, as rated both by themselves and by their friends and family.[17] So it seems psychedelics fully merit their classification as 'entheogens' – substances that 'generate the divine within' (from the Ancient Greek *entheos*, meaning 'filled with or inspired by god', and *genesthai*, 'generating').

At the time of writing, a study is under way led by William Richards at Johns Hopkins University in Baltimore, US, in which Catholic, Orthodox and Presbyterian priests, a Zen Buddhist and several rabbis have been given high doses of psilocybin to discover whether there are any persisting changes in their work and religious attitudes a year later.[18] Meanwhile, other studies add to evidence that mystical experiences during psychedelic trips have long-term benefits both for clinical and non-clinical populations that go beyond the purely religious or spiritual. Experiences of this kind, perhaps by challenging long-established mental habits, may help smokers cut down or give up completely, for example, and ease the anxiety and depression associated with life-threatening illnesses. Among healthy people, having such an experience may lead to improvements in the beneficial personality trait of openness that last for at least a year.[19–21]

If psychotherapists want to ensure the best outcomes for their patients in psychedelic clinics of the future, they may have to lay aside any misgivings they might have about crystals, crucifixes and the like. Such symbolic, religious or spiritual objects – perhaps personal items brought into the clinic by patients themselves – will be invaluable for creating the optimal set and setting for healing to occur.

The defining feature of any mystical experience, religious or secular, chemically-induced or the result of spiritual practices such as fasting or prayer, is the transcendence of our familiar sense of selfhood or ego. Ordinarily, every experience is filtered through thick, distorting goggles that tell us 'this is my body, my story, how I feel, what I want, what is mine'. Only when our ego goggles have been whisked away, by whatever means, can the universal mystical experience of oneness with nature, humanity, the universe or the divine take place. That psychedelic drugs do this 'to order' raises the tantalizing prospect of pinpointing exactly how the brain creates our selfhood, the defining feature of everyday consciousness and what we naturally assume ourselves to *be*. Already we can say with some confidence that a sense of having a discrete ego or self isn't necessary for consciousness, otherwise how could a psychonaut remain fully aware after losing it? Those unfortunate enough to suffer from depersonalization disorder, characterized by long periods in which they have no sense of subjective selfhood, will also testify to this truth.

You may have got the impression from the past couple of chapters that imaging studies of volunteers on psychedelics have already located the brain's ego goggles. Early research suggested that disruption of activity in the posterior cingulate cortex (PCC), the key hub of the DMN, was solely responsible for ego dissolution. But the self has many faces. In 2015, when Lebedev and the Imperial psychedelic researchers scanned the brains of people on a low dose of psilocybin, they found that their reports of ego dissolution were associated not with disintegration of the DMN, but a completely different set of regions known as the 'salience network'.[22] This network's job is to draw our attention to the most important features of the environment – not only potentially rewarding features, such as sources of food or potential mates, but also threats, such as predators or oncoming traffic.

As such, while the DMN is vital for autobiographical or narrative selfhood, the salience network seems to underlie our sense of being an agent with biological needs. When we survey the world, it's as if a map colour-coded for personal salience or *meaningfulness* is overlaid on everything we see, hear, touch, taste and smell, highlighting the features the brain predicts will be most important for our survival (and that of our genes). Under the influence of a psychedelic, however, this map of everyday salience is stripped away, allowing our senses the freedom to range more widely, making previously uninteresting stimuli seem more personally relevant.

One of the wonders of tripping on psilocybin with your eyes wide open is how extraordinarily meaningful perception becomes, as I found during the experience outlined at the start of this chapter. I'll never forget sitting under a young oak tree on the shady edge of our Dutch garden and, looking up, first falling under the spell of the sun striking through its leaves, then the swallows darting low over the lawn hunting insects. Later I became just as absorbed by the ingredients listed on the side of a packet of chewing gum. (I found it laugh-out-loud funny that Wrigley's feels it necessary, after listing the chemical constituents of its gum, to remind consumers that 'a varied and balanced diet and a healthy lifestyle are important'.)

In addition to revealing the importance of the salience network for this aspect of selfhood, Lebedev showed that psilocybin decreased connectivity between cortical regions and the medial temporal lobes: the core of the brain's prediction processing hierarchies. One of the salience network's two principal nodes is the dorsal anterior cingulate cortex, a region of cortex on the inner faces of the hemispheres that curls around the bundle of fibres connecting them. This registers 'surprise' when top-down predictions originating in the temporal lobes fail to explain away incoming sensory data. The other principal node of the salience network is the anterior insula, which is associated with conscious awareness of feelings and emotional states. Both regions are peppered with the serotonin-2A receptors to which psychedelic molecules such as psilocybin bind. By binding to these receptors, psilocybin seems to disturb the integrity of the salience network and reduce the top-down supremacy of the temporal lobes, scrambling established patterns of emotional salience.

Psilocybin's effect on salience goes some way towards explaining its paradoxical ability to make desirable but commonplace things like food more pleasurable, while taking away much of one's appetite. It also helps explain the drug's potential for treating addiction, which arises when particular features of the environment, such as alcoholic drinks, cigarettes, or even objects of sexual desire, become excessively, irresistibly salient. On psilocybin one can range impartially over the whole menu, rather than focusing on the dishes you have been conditioned through experience to favour. Metaphorically, the addict is obsessively drawn to ice cream and profiteroles and no longer derives any pleasure from anything else. Psychedelic therapy may return his or her attention to healthier options by reducing the intense salience of unhealthy favourites. To return to the rubber-sheet analogy introduced in chapter 5 (see page 128), psychedelics flatten the mindscape of desire, making the deeper troughs less treacherous for the rolling marble of consciousness and increasing the likelihood that safer, shallower troughs will be visited.

The salience network helps give rise to the kind of 'I' that wants. While low doses of psilocybin seem to interfere with this dimension of selfhood, other aspects, including the narrative or autobiographical self, sense of agency, embodiment and location in time and space, may only topple once the concentration of psilocybin in the bloodstream has reached a critical threshold, triggering a 'peak' or mystical experience. It's worth remembering that these are the same dissociable aspects of selfhood that hypnosis, trance, bodily illusions and virtual reality tease apart. In ordinary consciousness, these independent components of 'I' are somehow bundled together – perhaps through the low-frequency, alpha oscillations that psychedelics disrupt – to give the pervasive sense of having a unified, solid self that persists unchanging through time.

Like all the contents of consciousness, however, on closer inspection what we consider to be our true self turns out to be a convenient fiction. The feeling of having a unified self is an informed guess based on a history of conditioning stretching all the way back to the womb, a prediction that the brain makes about interactions between it and everything else. As we've seen again and again, selfhood has several independent dimensions, each a model arising from what happens when we act upon a particular facet of our world. When Descartes said 'I think, therefore I am', he only had part of the picture. Neuroscience can now fill in some of the blanks:

> This experience is emotionally salient, therefore I am,
> My five senses locate me here, therefore I am,
> I make things happen in the inner world of my body,
> therefore I am,
> I make things happen in the outside world, therefore I am,
> I remember the story of my life, therefore I am.
> I think, therefore I am.

Just for a moment, when we awaken to reality from a convincing dream, when we fall under the spell of a stage hypnotist, wear a virtual-reality headset or are fooled into thinking a rubber hand is part of our body, the curtain is drawn aside to reveal one aspect of the truth. But these are parlour tricks compared with the insights afforded by high doses of psychedelic drugs that *none* of our selves is solid and enduring. The revelation can be so transformative we call it 'mystical'. Chris Letheby and Philip Gerrans, philosophers of cognitive science at the University of Adelaide in Australia, write that it's the 'existential shock' of discovering we are not the thing we have always assumed ourselves to be that makes peak psychedelic experiences so therapeutic.[23] 'The subsequent diminution in the sense of solid selfhood shows subjects that this sense is ultimately just one more conscious experience, rather than a transcendental precondition of all such experiences.'

If you have never experienced it for yourself, you may be wondering what drug-induced ego death (DIED in the jargon) feels like. The experience can only really be hinted at. During my Dutch psilocybin adventure, escaping to my room out of the fierce heat and sunlight of the garden for a few minutes, when I tidied away a worn denim jacket in a cupboard and closed the door I got the distinct impression it was the old *me* hanging there in the cupboard. A few days after the psychedelic weekend, I had to take my car into the garage for a service. Within hours I received an automated text message from my insurance provider which, as a condition of my contract, had fitted a satellite tracker in the car that records speed and mileage. The message read: 'Hi, this is a service notification to inform you that we have detected your battery has been disconnected or has become flat in vehicle ER15 UJK'. DIED felt a bit like that. The bodywork remained intact, but James Kingsland's battery had been disconnected.

Disconnection has been a recurring theme throughout this book. According to the prediction error hypothesis of brain operation, like my car at the garage during its service, disconnection is sometimes necessary for routine maintenance – to regain flexibility or plasticity (change the oil, the filters and spark plugs). It almost certainly happens every night in our dreams, and is gloriously evident in the changes in perception wrought by psychedelic drugs. In the introduction to this chapter I mentioned that, while my fellow psychonauts and I were tripping inside the ceremonial hall, outside the garden was being magically transformed into Eden. When I walked out onto the lawn and looked around me, the colours were a thousand times more vivid than they had been before, and everything – the trees, the grass, the flowers, the birds – looked stunningly three-dimensional and gorgeously alien, like the plants and creatures on the lush green planet Pandora in the 3-D film *Avatar*. 'The world was as if newly created,' Hofmann wrote of the morning afterglow following his frightening LSD trip. I finally knew what he meant.

On psychedelics it's as if you are looking at the world through a child's eyes once more, because what the drugs seem to do is reduce the influence of long-established, top-down predictions or 'prior beliefs' about how the world should operate and appear. Once again, there are strong parallels with psychosis, a condition in which control over the relative influence of descending predictions and ascending sensory prediction errors is also disrupted. The changes in perception wrought by psychedelics and those associated with psychosis have certain important characteristics in common, as revealed by the 'hollow mask illusion'. In this well-known visual effect, a revolving mask depicting a human face, rather than alternating between the appearance of a convex, solid face and a hollowed-out, concave mask, continues to look convex and face-like even as the hollow inside of the mask rotates into view (you can watch an animation of the illusion online at https://bit.ly/2z4qKKp). People with schizophrenia aren't fooled by the illusion, however, and neither are people on LSD.[24] They see the mask first as a convex face and then as a hollow shell – in other words they see it for what it actually is.

But surely psychedelics and psychosis *distort* reality? Of course, in many ways they do, for example both can cause hallucinations and delusions. What these altered states reveal is the delicate balance the brain must continually strike between its prior beliefs about the world and sensory evidence that is often noisy and unreliable. From the cradle onwards, one of the multiplicity of things our visual system learns is that human faces are solid, convex objects. This becomes the brain's default interpretation, especially when the sensory data are ambiguous. But like all perception, 'faces are convex' is merely a hypothesis, albeit one based on long experience that is nearly always trustworthy.

Prior beliefs such as this about the regularities in our environment are weakened in schizophrenia patients and psychedelic trippers, whose brains invest less trust in 'received wisdom' about how the world should look and more faith in the raw sensory material. The same principle can be invoked to explain why on a bright sunny day, under the influence of a psychedelic, colours are wonderfully vivid and everything looks fantastically, preternaturally three-dimensional. Rather than being distortions of reality, these perceptions might actually be closer to the truth: less reliant on jaded preconceptions about how, for example, a tree should look and more on the streams of photons bouncing off the tree itself and striking the retina. It takes the genius of a great artist, perhaps a deeply troubled genius like Vincent van Gogh, to capture this vibrant, raw essence so that anyone can see it for themselves. It may be for our own good that the healthy brain filters perception through the thick lens of experience in ordinary waking consciousness. One can only imagine how overwhelming it would be to see the world like this *all the time.*

The temporary revolution in perception afforded by psychedelics can have long-lasting effects. There are even anecdotal reports of colour-blind men seeing colours for the first time. David Nutt at Imperial College London has been collecting some of them:

> One guy wrote to us to say 'I've been colour-blind all my life. My brother is an art critic and I've never understood what the hell he's talking about. I had mushrooms with him last weekend, I went to the Tate and for the first time I understood art because I could see colours!'

The effects can apparently last several months after a single dose of psilocybin, Nutt told me. Many neuroscientists remain sceptical, insisting that the root cause of colour-blindness is physical deficiencies in retinal cones. But what is 'colour' anyway? As I hope this book has demonstrated, the idea that conscious perception is a one-way street from sense organ to cortex is looking increasingly outdated.

Even a diehard cynic would have trouble disputing the acute changes in perception and selfhood wrought by psychedelic drugs and their associated spiritual or mystical effects, as I can now testify. But as one of our guardian angels advised us, in some ways the trip is the easy bit. The hard work starts when you try to integrate the lessons you have learned into ordinary life, which is the principal topic of the next chapter.

'Psychedelics allow you briefly to hear your personal language of subjectivity as sound, not meaning,' wrote the philosophers Gerrans and Letheby in a 2017 essay for the online magazine *Aeon* about the therapeutic potential of psychedelic-induced ego dissolution.[25] 'Whether you want to learn another language of selfhood is up to you.'

# 8

# *The Wonderful Lightness of Being*

Albert Hofmann, the creator of LSD, became convinced that psychedelics and meditation have a complementary effect on the mind. On the face of it this seems like wishful thinking. What could an ancient spiritual discipline possibly have in common with mind-bending chemicals? An intensive meditation retreat, free from all worldly distractions, seemed like the perfect place to put Hofmann's idea to the test, though the chilly shrine room of the English monastery in the Chiltern Hills where thirty or so of us perched cross-legged on cushions trying to focus on our breath seemed a million miles away from the sweltering maloca in Peru where – just a few months earlier – I had sat on a plastic mattress waiting anxiously to meet Mother Ayahuasca. The chorus of barks, croaks and growls on every side did remind me of that night in the jungle, though here the source was the noisy plumbing of the human digestive system rather than prowling nocturnal creatures.

On the first day of the autumn retreat at Amaravati Buddhist Monastery, an outpost of the ascetic Thai Forest Tradition, one of the eight precepts we undertook to observe for the next ten days was to abstain from consuming intoxicating drugs of any kind. The irony of abstinence being imposed by a generation of monks who freely admit experimenting with psychedelics and other psychoactive substances in the sixties wasn't lost on me. I know of several children of the revolution who found their way to Amaravati and its sister monasteries across the world. Many Western Zen and Tibetan Buddhists of that generation also decry any drug that leads to 'heedlessness', despite many admitting to using these substances freely in their own youth.[1]

On the evening of the second day of our retreat, Ajahn Vimalo, the seventy-two-year-old English monk in charge, gave a Dhamma talk in which he reminisced about his own first meditation retreat in his twenties. It was, he said mischievously, 'like being on magic mushrooms'. The next one didn't go nearly so well – he'd had enough of sitting painfully in silence after just a few days – which does highlight a crucial difference between the dependable, flashbulb moments of realization afforded by high doses of psychedelic drugs and the twisting, rocky road to Nibbana that is mindfulness practice. Despite the glimpses of profound calm, joy and happiness that even beginners often experience, meditation can also be painful, dull and frustrating, and of course it can take a lifetime (or several) to attain enlightenment, transcending forever anger, greed, fear, agitation and other negative emotions that characterize ordinary consciousness. 'When I came to meditation, I thought, I've been through the hippie culture and experimented with certain things, I'll crack this within a couple of years if I put my mind to it!' Vimalo told the retreatants of all ages gathered in the shrine room that evening to hear his teaching. 'I've been out in space, had a few experiences, so half my brain cells have gone already – I've just got to work on the other half!'

He hastened to add that he hadn't been nearly as steeped in the sixties drug culture as some of the people he knew: 'I didn't get too much into all that stuff, though I got into it enough to confirm certain ideas I'd had. Then from the age of twenty-four I was a devout yogi and then a Buddhist.' He had friends who were pretty wild, he said. 'They used to say, you don't have to go *this* crazy!' But he told them he had to do it properly. Which was how, as if reborn onto a new plane of existence, he now found himself sitting in the lotus position beneath a statue of the Buddha, shaven-headed and wrapped in a monk's ochre robes, joking self-deprecatingly about the path he was still treading some five decades later. His early steps on this journey hadn't taken him straight to the monastery gates, however. First he would marry and raise two kids before finally renouncing worldly things at the age of forty-five and 'going forth into the homeless life', as Theravada Buddhists put it.

In all honesty, neither meditation nor a bitter draught of ayahuasca tea – nor for that matter a chewy mouthful of magic truffles – can provide a quick-and-easy fix for the human predicament. More often than not the trip or retreat is just the start. The hard work starts after the vivid afterglow has dimmed and you attempt to integrate any lessons you have learned into the messy, complicated business of everyday life with all its frustration, stress and temptation.

For some, it's true, the body and its appetites do seem to have changed irrevocably. At the Dios Ayahuasca Sanaciones healing centre in the Peruvian jungle near Iquitos, I met a man who told me that the tea had saved him from the clutches of addiction. He was working for a company in the City of London and had been an enthusiastic participant in its cocaine- and booze-fuelled culture, but ever since he first consumed ayahuasca in a series of ceremonies the previous year, drink and drugs no longer appealed to him. Now the merest whiff of the alcohol in someone else's drink at a party gave him a headache.

Nonetheless, for most people the ceremony only marks the beginning of the healing journey. In the past, in cultures where psychedelics such as ayahuasca were inseparable from religious practice, your community facilitated this process of integration. 'You were part of a group that had that experience together – you and the shaman would drink – so in a sense it was already integrated with your life,' says Gerald Thomas, a Canadian psychologist and ayahuasca researcher based at the University of Victoria, himself a veteran of half a dozen ceremonies. 'The community was involved and helped integrate and facilitate healing.' Our atomized, long-distance lifestyles have denied most people this continuing mutual support after a shared psychedelic experience, Thomas explained to me over a glitchy Skype connection from British Columbia. 'We parachute in for a weekend, we drink and we go back to our normal lives.'

Observational studies of modern ayahuasca churches seem to bear this out. They suggest that among those with former problematic drug use, being part of a community who regularly drink the brew together helps with integration. A 2018 survey of 1947 members of the União do Vegetal (UDV) church in Brazil found that while their overall, lifetime use of tobacco and alcohol was higher than the population average, their current use was significantly lower. And the longer they'd been part of the church and the more ceremonies they had attended in the past year, the less alcohol and tobacco they were now likely to be consuming.[2] Similar results have been found in other surveys of drug use among members of Santo Daime and UDV churches in both Brazil and the US.[3,4]

Retrospective studies like these provide fascinating clues about the potential of psychedelics to combat addiction and how they might work, but they're not ideal. Prospective 'before and after' trials, while they usually involve many fewer people, are easier to interpret because there aren't so many hidden variables. Thomas, who works with the Centre for Addictions Research of British Columbia at the University of Victoria and is Director of Alcohol and Gambling Policy at the British Columbia Ministry of Health, conducted one of the first such studies to look into ayahuasca and addiction. In 2013, he and his colleagues published a pilot study involving twelve people with substance abuse problems which found reductions in alcohol, tobacco and cocaine consumption up to six months after they took part in two ceremonies and several group counselling sessions.[5] Follow-up interviews revealed that, above and beyond any purely physiological change in their addictive urges, participants attributed their progress to an increased connection with others and nature, and a more spiritual, mindful way of being.

A forty-one-year-old woman told the researchers:

> Before the ceremony I was struggling with my addiction – crack cocaine – for many years. And when I went to this retreat, it more or less helped me release the hurt and pain that I was carrying around and trying to bury that hurt and pain with drugs and alcohol. Ever since this retreat I've been clean and sober. So it had a major impact on my life in a positive way... My family is back in my life. My daughter is back at home. And we are getting closer and closer every day as time goes on.

Compared with depression and anxiety, however, progress in bringing psychedelic-assisted treatment for drug addiction into the mainstream has been very slow. A pilot study in 2015 found psilocybin significantly increased the likelihood of abstinence following psychotherapy for alcohol dependence, but there have been no recent trials of LSD for treating alcoholism.[6] A meta-analysis published in 2012 of randomized controlled trials in the sixties, when many clinics were offering LSD-assisted therapy, suggested impressive decreases in dependence among patients.[7] Due to the drugs' current legal status, however, it's impossible to know how many doctors are still quietly using psychedelics 'off label' to treat addiction, or for that matter how many patients are taking matters into their own hands, for example travelling to South America to take part in ayahuasca ceremonies.

Much more research remains to be done, but Thomas suspects that substance abusers' decreased reliance on drugs after taking ayahuasca may be as much about a greater capacity to accept and reconcile past traumatic events as it is about the brute biology of craving. As a result of this increased acceptance they become less likely to use alcohol or narcotics to numb or distract themselves from painful emotions and memories. His study and others suggest that psychedelics, at least in the short term, can improve one's capacity to be mindful – non-judgementally present in the moment and accepting of both pleasant and unpleasant emotions – a trait that meditation also aims to cultivate.[8–10] One of the objectives of integration following a psychedelic experience, therefore, is to nurture this new-found ability. 'It's about learning to be present in uncomfortable sensations,' says Thomas, 'allowing the sensations to arise in the body and feeling the fear, feeling the loss, feeling whatever it is that you've not been allowing yourself to feel. It won't necessarily be pleasant, but that's what it is to be human.'

Everyone who has had a psychedelic experience, not only those with substance abuse problems, could benefit from this ongoing integration process. According to research published in 2018 by Roland Griffiths and his colleagues at Johns Hopkins University School of Medicine in Baltimore, US, regular meditation, mindfulness in daily activities, keeping a journal and secular, 'spiritual' pursuits, such as spending time in nature, exploring your artistic side and volunteering, can work wonders for healthy people in the months after they take a high dose of psilocybin.

Griffiths and his team randomly assigned seventy-five volunteers to three groups of twenty-five each. The first received a very low dose of psilocybin and a moderate amount of support for integration practices; the second got a high dose and moderate support; and the third got a high dose and intensive support. Six months down the line, those in the high-dose, intensive support group were out in front on a whole range of psychological and social measures, such as closeness in their personal relationships, a sense of meaning and purpose in their lives, and transcendence of the fear of death. Importantly, evidence of personality changes came not just from the subjects themselves but also from friends, family and colleagues. Two factors seemed to mediate these beneficial effects: the intensity of any mystical experiences people had while on psilocybin, and their level of engagement with the spiritual practices that followed. Participants were encouraged to practise sitting meditation for between ten and thirty minutes every day, and of all the recommended activities designed to foster psychological and spiritual well-being, sticking to this target seemed to have the biggest impact on their well-being and outlook.[11]

It's starting to look as though meditation may be one of the best ways to integrate the extraordinary experience of a psychedelic trip into the ordinariness of everyday life. When researchers 'lift the bonnet' and peek inside people's heads as they meditate or are under the influence of a psychedelic – using techniques such as functional magnetic resonance imaging (fMRI) – they find that both pursuits have uncannily similar signatures in the brain.[12] It's as if meditation provokes a neural 'echo' of what happened during the trip, which may explain why it is such an effective tool for sustaining and enhancing any long-term benefits.

Of particular interest is the way both meditation and psychedelics disrupt the default mode network (DMN), which, as we saw in Chapter 5, fires up whenever the brain isn't occupied with performing a task requiring our undivided attention. Both meditation and psychedelics are known to decrease activity and connectivity within the 'medial' DMN, the components located in the brain's midline regions on the inward-facing surfaces of each hemisphere, namely the posterior cingulate cortex (PCC), parahippocampus and medial prefrontal cortex. These areas are involved in reviewing the past, planning the future and mentalizing or 'theory of mind' – simulating others' perspectives and intuiting their thoughts, feelings and intentions. They are also closely associated with the worries, rumination and self-related judgements that can arise spontaneously when our minds wander from the task at hand.

When people practise 'focused attention' meditation – concentrating on a discrete bodily sensation such as the breath and gently bringing their attention back whenever they notice that the mind has wandered – their sense of social, autobiographical selfhood or ego diminishes as activity in the medial DMN is suppressed. In effect, meditators are taking advantage of the way our brains organize their processing power, see-sawing between tasks requiring focused attention (the job of its salience and attention networks) and internally focused cognition (the responsibility of the DMN). When the task-focused networks are busy, activity in the DMN is automatically suppressed, in much the same way many companies now ban their staff from using social networks like Instagram and Twitter during work hours.

By keeping the brain's task-focused networks occupied, focused-attention meditation suppresses chatter within its built-in social networks. This work is effectively done for us when psychedelic molecules bind to the serotonin 2A receptors found in abundance in the PCC, parahippocampus and medial prefrontal cortex of our brains, hampering their neurons' ability to fire in synchrony. As we've seen, in ordinary consciousness the PCC in particular acts like a conductor, broadcasting low-frequency electrical signals, known as alpha waves, that orchestrate the brain's top-down perceptual and cognitive activities. But whereas psychedelics short circuit the PCC by binding to 2A receptors and preventing it from generating these signals, meditation gently dials down its activity by diverting our attentional resources elsewhere.

Overactivity in the DMN in general and the PCC in particular has been linked to the excessive worry and self-critical rumination that characterize anxiety and depression. So it figures that both meditation and psychedelics may be highly effective approaches for combating these conditions. In the eighties, psychologists developed a meditation-based treatment called mindfulness-based cognitive therapy (MBCT) which, in several large clinical trials, has proved its worth for preventing relapse in people who have experienced several episodes of major depression. Psilocybin, meanwhile, began its own journey towards mainstream medical acceptance as a depression treatment in a pilot study published in 2016, and is now being put through its paces in larger clinical trials.[13,14]

By pulling the plug on the PCC in the medial DMN (or, to stretch my earlier analogy, preventing access to Twitter), psychedelics or the focused attention of an experienced meditator can precipitate ego dissolution. Reducing activity and connectivity in the *lateral* DMN on the outward-facing surfaces of each hemisphere, meanwhile, scuppers another facet of selfhood: our feeling of being located at a particular point in time and space.

The temporoparietal junction or TPJ is an important region

within the lateral DMN. We first encountered it in Chapter 3 in the context of virtual reality and out-of-body experiences (see page 75). From its vantage point at the intersection of the temporal and parietal lobes, the TPJ integrates sensory information to create something like a 'You are here' arrow on a map (or the 'I'm here!' of a picture uploaded to Instagram). This arrow of selfhood tells us not only where we are in relation to everything in our immediate environment, but also provides a timeline from which we can infer the order in which conscious events have been happening – a vital clue whenever the brain needs to predict what caused what. Disrupting the TPJ's activity and connectivity, either through meditative absorption or serotonin 2A binding, may give rise to the mystical experiences of timelessness and spacelessness (contributing to 'oceanic' feelings of oneness with the universe) that both seasoned meditators and people on psychedelics report. It's another way in which both practices dissolve the distinctions that we learned from our mother's breast onwards between effects and their causes, between ourselves and others – between us and everything else.

The shared key that unlocks the psychological benefits of both psychedelic drugs and meditation may be the way the brain creates these different dimensions of selfhood. According to Buddhist teaching, the ingrained belief that residing within each of us is a solid, indivisible, unchanging 'self' – a self that is somehow the measure of the entire world – is a delusion that leads to suffering. The idea is that by dispelling this pervasive perceptual illusion, through the dogged practice of meditation and mindfulness and by adhering to high standards of ethical behaviour, we can transcend most if not all our suffering. This may go a long way towards explaining why so many young people who dabbled in psychedelics in the sixties would gravitate towards Buddhism in later life: serotonin 2A agonists had given them a foretaste of Nibbana. Taken under appropriate conditions and in sensible amounts, LSD, mescaline and psilocybin provide a glimpse of what a life free of the heavy chains of selfhood might feel like.

One of the mystifying things about meditation that most people can't get their heads around when they first start practising is that it only works its wonders when you allow yourself to *do nothing*. In a guided meditation session, the teacher will often intone something like: 'We're not going anywhere, not doing anything, not trying to change anything, just focusing on our breath.' I for one found this confusing when I began meditating around ten years ago with the attitude that, by hook or by crook, I was going to 'fix' my wonky mind. And yet Buddhist teachers and secular mindfulness therapists kept insisting there was nothing wrong with my mind, or for that matter anybody else's. There was nothing that needed changing. This didn't make any sense to me. The psychotherapists said they were using mindfulness and meditation to treat (or fix, as I saw it) anxiety and depression, and one of my favourite sayings of the Buddha was that 'our life is shaped by our mind; we become what we think. Suffering follows an evil thought as the wheels of a cart follow the oxen that draw it.'[15] How could anyone hope to alleviate suffering, therefore, without first changing their mind?

What I had failed to grasp was the distinction between our attitude to ongoing conscious experience and abstract aspirations for the future. Fully accepting the way things are in this moment is a mindset that doesn't preclude having long-term goals – even monks have those. But to dispel the illusion of selfhood without the help of chemicals one must first cultivate a mental attitude of 'doing and changing nothing'. On this, neuroscientists and Buddhists are agreed: the moment-by-moment experience of selfhood arises when we are striving to change our world in some way.

According to the prediction error theory of brain function, our emotions and every dimension of selfhood are the consequences of physical, cognitive or physiological action. To recap, the theory proposes that the brain is a prediction machine for guessing the hidden causes of its sensory inputs. Its predictive models of how the world works comprise multilayered information-processing hierarchies stretching from the brain's executive regions in the prefrontal cortex all the way down to sensory receptors in our sense organs, muscles and internal organs. The predictions that the models continually make about the causes of sensory stimuli are dispatched from the prefrontal cortex to the sensors, while error messages are pinged back in the opposite direction. To arrive at a sufficiently accurate model of reality, each intermediate layer of the hierarchy must somehow minimize these prediction errors – the discrepancies between descending predictions and ascending sensory information.

As we saw in Chapter 4, there are two ways the errors can be resolved. They can either be allowed to pass upwards through the hierarchy so that our high-level predictions are updated to more closely match incoming sensory data – which is the essence of perception and learning – or the sample of sensory inputs at the lowest levels of the hierarchy can be changed to match the predictions, which is the essence of action (see Figure 2, page 45). In other words, we can either change our predictive models or seek out new sensory data that will confirm them. The latter approach is known as 'active inference' and it's how we control our muscles: the brain first predicts the desired movement's sensory consequences (the signals that will come from stretch receptors in the relevant muscles) then spinal reflexes automatically fulfil these predictions. This kind of active inference leads to a spin-off prediction, that of personal agency – 'I probably did that' – whenever we distinguish the sensory results of our own actions from those caused by external forces.

The central nervous system is also thought to regulate our physiology through active inference of interoceptive (internally generated) sensory inputs, keeping parameters such as body temperature, blood glucose and blood pressure at optimal levels. Here, the top-down predictions are enacted by the autonomic reflexes responsible for homeostasis. Exerting this kind of control gives rise to another facet of selfhood, namely presence or embodiment – 'This must be my body'. It is also bound up with our emotions, which according to the theory are our inferences about the causes of physiological changes.[16]

So, we use active inference to control our internal and external environments and in so doing we get a sense of being emotional, embodied agents acting upon the world. But the crucial thing to understand about active inference is that it can only do its work if sensory prediction errors are briefly prevented from passing all the way upwards through the processing hierarchies and updating our models of reality – which would give rise to conscious perception and learning rather than action. Ascending sensory prediction errors must be temporarily stifled, allowing spinal and autonomic reflexes to fulfil high-level predictions. By analogy, our CEO doesn't need to know or influence everything that goes on in her company. Indeed, where everyday tasks are concerned it's far better that staff with the necessary expertise working on the shop floor are allowed to get on with what they do best.

Imagine, for example, deciding to shake the hand of someone you've just met. According to the theory of active inference, you initiate this task by subconsciously predicting the sensory consequences of lifting your hand. But if instead of allowing your spinal reflexes to fulfil the prediction, sensory error messages from the stretch receptors in your muscles were to pass upwards through the predictive hierarchy, insisting confidently that your arm was hanging at your side, that is where it would stay. Moving your arm requires an unconscious, temporary suspension of attention to current sensory signals – a strategic downgrading of the confidence or 'precision' afforded to the associated prediction errors.

Neuroscientists have only just begun pondering how the most abstract forms of cognition, such as theory of mind, setting goals for the future and reflecting on the past – all skills under the dominion of the DMN – fit into this picture. It seems likely, however, that to intuit what someone else is doing, feeling or thinking, or to simulate mentally our own past and future actions, we deploy the very same active inference machinery that we would use to perform these actions ourselves in the here and now.[17,18] As with dreaming or playing a video game, the action only happens in our heads – it is 'virtual' or offline – but like online active inference it gives rise to feelings of selfhood and otherhood. The only practical distinction is that to avoid 'acting out' our simulations there must be a neuronal disconnect, a temporary roadblock, somewhere lower down in the predictive hierarchy.

Meditation – sitting motionless with an attitude of 'not going anywhere, not doing anything, not trying to fix anything' – is the polar opposite of active inference. If you're practising well, not only are you not moving a muscle, you're not making any plans to do so in the future. You are not reflecting upon and judging the past actions of you or anyone else, or simulating future actions. Meditation is all about immobility, virtual and actual. Even the pain in your back and knees leaves you emotionally unmoved. You are no longer inwardly complaining and have no desire to change anything. It may be no coincidence that this is the same accepting attitude that people about to take a psychedelic drug are encouraged to adopt. 'Don't fight it, just go with the flow,' they're told. 'If the experience is frightening or unpleasant, don't resist. Bear with it.'

This is being not doing. Non-judgemental, focused attention on the sensations associated with breathing, for example, reorients the brain's predictive machinery away from top-down, active inference towards the upward flow of this narrow subset of sensory prediction errors (it increases their relative precision). Once the mind has settled, the meditator can broaden his or her mindful attention to encompass anything else that impinges on consciousness (a technique known as 'open monitoring'). But regardless of the focus, broad or narrow, rather than the mind initiating action of any kind, sensory stimuli are allowed to continually update ongoing predictive models of reality. Active inference has been replaced by radical acceptance: 'This is simply how things are in this moment. There's nothing wrong here, nothing that needs changing.' The idea is that, with sustained practice, this mindful outlook will start to seep into ordinary consciousness. Some studies suggest it can start to happen in as little as eight weeks – the length of a mindfulness-based stress reduction (MBSR) course – though the full benefits may take years to be realized.

During meditation, by reducing active inference, sustained mindful attention diminishes the aspects of selfhood that normally arise from it, including personal agency, autobiographical, emotional selfhood and embodiment. The practice suppresses activity high in the brain's predictive hierarchies, for example within the DMN, which normally plays a prominent role – via active inference – not only in theory of mind, reminiscing about the past and planning the future, but also our feeling of autobiographical selfhood. In effect, the tide turns: the usual top-down stream of active inference ebbs and sensory prediction errors are allowed to flow all the way up through the hierarchy to update our predictive models.

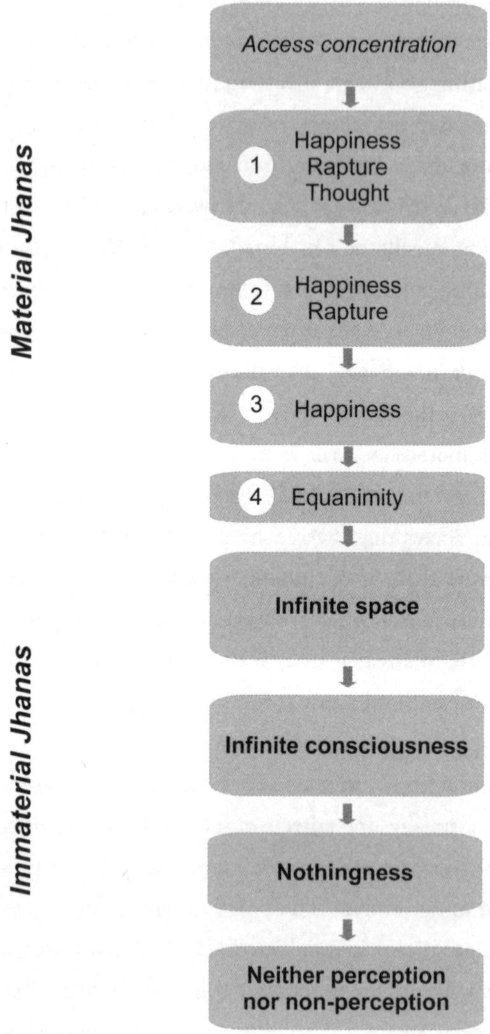

*Access concentration*

**Material Jhanas**

1 Happiness
Rapture
Thought

2 Happiness
Rapture

3 Happiness

4 Equanimity

**Immaterial Jhanas**

**Infinite space**

**Infinite consciousness**

**Nothingness**

**Neither perception nor non-perception**

Figure 7: *Absorption allows meditators to experience eight progressively deeper, more abstract jhana states of consciousness*

All manner of weird and trippy things start to happen when the mind is relieved of the burden of active inference. As Theravada Buddhists describe it, by focusing exclusively on a sensation such as the breath and letting go of everything else, an adept meditator attains 'access concentration' before sinking into eight progressively deeper levels of absorption called the *jhanas* (see Figure 7). In the first four, called the 'material' or *rupa jhanas*, awareness of the external world diminishes and the sensations of breathing become increasingly subtle, until finally the breath seems to disappear completely. Intense feelings of bliss then arise, which are later replaced by pure happiness and then perfect equanimity. Any wispy thoughts that remain evaporate like mist. Words are inadequate to describe the four 'immaterial' or *arupa jhanas* that follow, but in an echo of mystical experiences during a psychedelic trip, the rare individuals who can access these four extraordinary altered states through meditation label them 'infinite awareness', 'nothingness' and 'neither perception nor non-perception'.[19] Unlike the preceding material jhanas, the meditator has now passed beyond sensory awareness of any kind.

So with each successive jhana, consciousness becomes increasingly removed from external sensory awareness. According to Buddhist philosophy, the eight jhanas are equivalent to the four lower (physical) and four higher (heavenly) spheres of existence into which beings can be reborn, dependent on the kamma they accumulated in previous lives.

In the first immaterial jhana, writes the Sri Lankan Buddhist monk Henepola Gunaratana, the object of attention is not the contents of the mind but the space they occupy.[20] 'The mind as a space, medium, channel, or vehicle is your object of awareness,' he writes. In the second, contemplation of space is replaced with contemplation of awareness itself. 'You dwell in boundless consciousness,' he says, 'pure awareness of awareness.' In the third immaterial jhana, the meditator contemplates the lack of any object of any kind – nothingness or the void. Even this is a kind of perception, however, which must be abandoned. 'Total absence of perception is sublime,' says Gunaratana of the final immaterial jhana.

The mundane activity of focusing one's attention on a narrow sensory input such as the breath has precipitated a total collapse in selfhood and perception. A seminal paper published almost three decades ago hinted at what it would be like to be a perfectly enlightened human prediction machine. The Harvard mathematician David Mumford, a pioneer of the prediction error theory of brain function, speculated in 1992 that in an 'ultimate stable state' nerve cells delivering predictions downwards through the processing hierarchy would perfectly predict what each lower level was reporting back, with the result that lower levels responsible for sending sensory signals (prediction errors) upwards would cease firing at all.[21] In an intriguing footnote he wrote: 'In some sense, this is the state that the cortex is trying to achieve: perfect prediction of the world, like the oriental Nirvana, as Tai-Sing Lee [his student and collaborator] suggested to me, when nothing surprises you and new stimuli cause the merest ripple in your consciousness.'

Viewed in this light, jhana meditation adds further evidence for the prediction error theory of perception. Alongside other altered states such as dreams, trance and hypnosis – not to mention sensory illusions and the placebo effect – meditative absorption confronts us with the disorientating possibility that everything we perceive is little more than a hallucinatory hypothesis. Because when the brain predicts with absolute precision and accuracy an ongoing sensory input, such as how it feels as the breath enters and leaves the body, these sensations vanish as if they were never really there in the first place.

And yet, unlike deep sleep or anaesthesia, in jhana states conscious awareness is still very much present, and it's a particularly pleasant kind of consciousness. Meditators describe feelings of intense joy and happiness as they experience the first two rupa (material) jhanas. Where do these emotions come from? In 2013, researchers at the University of California and the University of Washington used fMRI and electroencephalography (EEG) to monitor the brain activity of a seasoned meditator as he experienced the first jhana (which he compared to 'continuous multiple orgasms') and the second (which he said was like 'opening a birthday gift and getting exactly what you most wished for'). The scientists found evidence that the second jhana activated his brain's dopamine reward system (unfortunately they were unable to obtain reliable scanner readings during the first jhana because his head moved). This was a first, because his dopamine system appeared to have been activated in the absence of any rewarding, external stimulus. His brain was getting a kick not from the usual things, such as sex, fast cars or chocolate, but from meditation.[22]

In common with drugs of abuse such as cocaine, jhana meditation appears to hijack our reward circuits. And yet the underlying neurochemistry must be very different, because unlike a drug, the practice doesn't cause dependence or withdrawal symptoms, the classic signs of addiction. Furthermore, Buddhists don't view these powerful sensations of pleasure and happiness as an end in themselves. Rather, they are seen as an incidental side effect of developing the concentrative skills needed to penetrate the true nature of all conscious phenomena – the tools that can help them attain spiritual enlightenment.

It would be rash to jump to any firm conclusions about the

neuroscience of jhana meditation based on a single study involving a single subject, but the way dopamine neurons signal pleasure and displeasure is scientifically well established. These nerve cells report not the size of a rewarding stimulus but rather a 'reward prediction error': the discrepancy between the reward we expected to get and what we actually receive.[23]

This is fundamental to how we learn. If a particular behaviour elicits less of a reward than predicted, dopamine neuron activity decreases. There's a feeling of disappointment and the behaviour becomes less likely in future. If, on the other hand, it elicits a bigger reward than expected, there is a spike in the activity of our dopamine neurons and we experience pleasure. As a result, the prediction is upgraded to reflect the size of the reward and we'll be more likely to perform the associated behaviour in future. But here's the bad news. Now that we're getting exactly what we predicted, dopamine neuron activity will remain unchanged at baseline levels. The behaviour isn't as rewarding as it was the very first time. This set-up goes a long way towards explaining our innate greed and vulnerability to addiction, because in order to perpetuate a dopamine buzz we must continually seek out increasingly large or novel rewards. What we've got will never be enough: we are wired to want more and more and more. Like problem gamblers, we can't help ourselves.

Have monastics found a way to 'beat the house', in casino parlance, enjoying the thrill of a dopamine high in the absence of any external stimulus and without the ruinous costs of addiction? You don't have to be a monk to reach these meditative states, writes Leigh Brasington, one of the authors of the jhana research. The secret is to expect nothing. Just focus on your breath. When that starts to disappear, switch all your attention to a pleasant sensation anywhere in the body and then do nothing else. 'The hard part is the "do nothing else" part,' he writes.[24] Brasington is not himself a scientist but a meditator of many years' experience who leads jhana retreats. 'Wait for the jhana to come find you.' When it does, he says, you'll know because that little pleasant sensation will blossom into all-suffusing ecstasy.

It's worth emphasizing here that jhana meditation, while in theory it can be practised by anyone with the requisite patience and persistence, is a rare accomplishment even among monastics. The Buddha himself is said to have used the practice to attain Nibbana or complete enlightenment as he sat meditating under the Bodhi tree. Some Buddhists therefore believe enlightenment can only be reached by those who have perfected the practice, though others beg to differ.

What is certain is that the jhanas are not awakening itself, but rather a skilful way to concentrate the mind on the path to awakening. From a purely scientific perspective they are a powerful demonstration of what happens when the precision of a narrow subset of sensory prediction errors is enhanced and the influence of active, top-down inferences is diminished. This, in a nutshell, is what this book has been all about, because what dreams, trance, hypnosis, psychedelics and meditation all have in common is that each, in its own way, breaks down established patterns of thought and behaviour: the descending predictions about the world that hold sway over more ordinary, waking consciousness. Altered states offer a means to reboot the brain's predictive hierarchies and break the power of entrenched patterns of thought and behaviour – our maladaptive habits, preconceptions and prejudices. These patterns are the work of the highly experienced but often inflexible 'autopilot' that normally guides us. The insight that Buddhism offers is that without the reality checks afforded by mindful attention, this automatic way of existence is at best no more than an uneasy dream. At worst it can become the waking nightmare of mental illness or addiction.

I must confess, in the seemingly endless hours I spent sitting on a cushion in the shrine room of Amaravati's retreat centre, nothing could have been further from my mind than the joy and happiness of jhana. There was too much pain to leave any space for bliss: pain in my knees, pain in my lower back, pain between my shoulder blades. By the evenings, sitting on the floor had become so unbearable I resorted to sitting in a chair during Ajahn Vimalo's nightly Dhamma talk. And yet he, at the age of seventy-two, still showed no sign of discomfort after spending almost two hours in the lotus position. Monks do feel pain, he assured us, it just doesn't bother them any more. They are, however, just as prone to the effects of stiffness and numbness when they do eventually stagger to their feet. He joked about a former abbot, a venerable gentleman even older than he, who had to hop from the temple on one foot after a particularly long session during which his other leg had gone to sleep.

During a retreat in 2015, the current abbot, Ajahn Amaro, explained that monks transcend pain not through masochism, but acceptance. He described this state as 'the heart not complaining, not criticizing, not waiting for it to be over, but just in that moment open and aware, open and accepting: "It's like this. This is the feeling."' He claimed that the same attitude can transcend the emotional pains of grief, anger, frustration or depression. 'It works in exactly the same way. We learn to relate to it with open-hearted and radical acceptance. You're not pretending that you like it or want it, but here it is, just another arising and ceasing, a pattern of consciousness coming and going and changing.'[25]

One evening, Ajahn Vimalo spoke movingly about mourning the sudden death of one of his sons. Even as he mourned, he said, his mind remained still. He told us the practice of observing his mind with openness and compassion for so many years had transcended even this profound suffering:

> Through observing I started to come to an understanding. Dare I say a certain wisdom occurred. It hasn't come through thinking or reading lots of books about Buddhism and intellectual understanding. It just comes through a way of observing until the perception starts to change. And when our perception changes our world changes.

Buddhists would have us believe that on the one hand if you focus mindfully on a pleasant sensation it will blossom into bliss, and on the other if you focus mindfully on even the most unpleasant sensation – the worst imaginable physical or emotional pain – its hold over you can be broken. You could be forgiven for thinking this all sounds a little too good to be true. But while research into the joys of jhana meditation remains very much in its infancy, there is ample evidence from clinical trials that more ordinary meditative states and mindfulness can help prevent relapses in people prone to recurrent depression, and preliminary evidence suggests they can reduce cravings associated with drug addiction. Meditation has also been shown to reduce the intensity and unpleasantness of pain, but it's thought to work through a completely different mechanism than any other kind of pain relief. Opioid painkillers, such as codeine and fentanyl, work by binding to the same receptors as the brain's own, endogenous opioids – the endorphins that flood the body when, for example, athletes push through the pain barrier triggering a 'runner's high'. The painkilling, euphoric effects of both artificial and endogenous opioids are wiped out by a drug called naloxone, which binds to the same receptors. Interestingly, naloxone also blocks placebo analgesia – the surprisingly powerful effects of dummy 'painkilling pills' – suggesting that the mere expectation of pain relief causes the release of endorphins. An active inference that we're going to feel less pain *makes it so* by releasing the body's own painkillers.[26]

But this isn't how meditation relieves pain. When scientists at the University of Oregon and Wake Forest School of Medicine used electrodes to deliver harmless but painful electric shocks to the middle fingers of thirty-two meditating volunteers – following an injection of saline solution or an injection of naloxone – something totally unexpected happened. All the subjects were experienced meditators with a proven ability to reduce the unpleasantness and intensity of this kind of pain by 15 per cent or more. Before the experiment the researchers had predicted that, compared with saline, naloxone would make the electric shocks more painful for the meditators by preventing their endorphins from binding to opioid receptors. In actual fact, naloxone *enhanced* the painkilling effects of meditation. If meditators were injected with the drug, the electric shocks were less painful than if they'd been injected with saline solution.[27]

This suggests that meditative pain relief doesn't involve endorphins. Unlike placebo analgesia, it doesn't work by increasing our expectation that the pain will go away, unconsciously releasing a flood of the body's own painkillers. Rather, in line with Ajahn Amaro's explanation, meditative pain relief works by promoting acceptance of the pain, 'not complaining, not criticizing, not waiting for it to be over, but just in that moment open and aware'. In the language of prediction error processing, meditation reduces the downwards flow of predictions about painful stimuli and boosts bottom-up sensory signalling. And by favouring acceptance over action, our model of the world is updated: 'It's like this.' The researchers are at a loss to explain why naloxone *boosts* meditative pain relief, but it's possible that by blocking the numbing effect of endorphins the drug facilitates the upwards flow of sensory prediction errors, enhancing acceptance.

Another line of research adds further proof of the analgesic power of acceptance. A neuroimaging study has found that, compared to novices, mindfulness practitioners were able to reduce the perceived unpleasantness of pain by an impressive 22 per cent on average, and that this was associated with decreased activity in their lateral prefrontal cortex, involved in top-down cognitive or 'executive' control. There was also increased activity in areas of the brain that process sensory signals, namely the posterior insula and the somatosensory cortex, part of the outer surface of the parietal lobe of each hemisphere.[28]

Taken together, these findings endorse the widely accepted view among doctors and psychologists that pain perception is as much an active, emotional process as a passive one. There's a lot more to it than simply registering raw sensory data. Just like emotions such as anxiety, anger and elation, pain is a cognitive inference or prediction about the causes of unexpected interoceptive stimuli. When people practise open monitoring meditation they bring mindful, non-judgemental attention to bear on bodily sensations such as pain wherever they arise, so it figures that this should be accompanied by increased activity in the insula (responsible for interoception) and decreased activity in the prefrontal cortex (responsible for top-down cognitive inferences).

Presumably the same mechanism underpins another proven benefit of long-term meditation practice: enhanced emotional regulation. This may be one more good reason to incorporate meditation into one's post-psychedelic integration routine. However, in common with other altered states of consciousness explored in this book, meditation is not without its risks for the unprepared or psychologically vulnerable, in particular those experiencing their first intensive retreat.

Monasteries have been described as beautiful prisons. Often located in stunning natural settings, they are havens of calm and tranquillity run in such a way that people can pursue their spiritual goals free from everyday distractions such as money, relationships and careers. Nonetheless, for the very same reasons, they are also socially isolating and psychologically challenging places to stay. On arrival at Amaravati, we were obliged to surrender our mobile phones, which were locked away in the office safe until the end of the retreat. Needless to say, the monastery offered no TV, radio or Internet as alternative mental diversions. For much of the time there was nothing to do and nowhere to go, which of course was the point. A monastery is not a hotel. Increasing the sense of isolation, reading of any kind was discouraged and talking forbidden, apart from essential communication during working meditation in the kitchen or grounds. Even eye contact became rare.

These restrictions help one develop the necessary mental discipline and concentration to practise mindfulness. Nonetheless, social isolation in unfamiliar surroundings among strangers can be difficult. It's also far from ideal if a psychological crisis should arise, because the usual sources of help and support from friends, family, partners or medical professionals are temporarily unavailable. In addition, at Amaravati most of us were accommodated in dormitories, so there was no private space to which we could escape.

Adverse reactions to meditation are extremely rare when it's part of a psychotherapy such as MBCT. But, perhaps as a result of the additional pressures outlined above, meditation on retreats has been known to precipitate mental health crises, including psychosis, depersonalization, out-of-body experiences and mania. This isn't surprising given the profound changes in consciousness associated with deep meditation, including emotional highs and alterations in perception, selfhood and embodiment. After a recent survey of Western retreat participants and teachers, the psychiatrist Willoughby Britton and her colleagues at Brown University Medical School in Providence, Rhode Island, wrote that some of the challenging experiences they uncovered resembled the classic symptoms of schizophrenia. Nonetheless, the researchers also acknowledged the difficulty of disentangling pathological from spiritual phenomena. They wrote that some Buddhists would interpret such experiences as signs of spiritual progress.[29]

Towards the end of the autumn retreat at Amaravati, one of my fellow retreatants, a man in his twenties relatively new to meditation, developed what a psychiatrist might describe as hypomania. He felt 'wired', his thoughts racing, and was unable to sleep. He said this followed a session during which the pain that had been troubling him for days suddenly 'switched off'. That evening he appealed to the monks leading the retreat for help, who counselled and supported him late into the night. He later told me that during the pre-dawn meditation session the next morning, having failed to get a wink of sleep, he was relieved to discover that the pain was back and, with its return, a sense of normality was restored.

Experiences like this serve as a reminder that altered states of consciousness, chemical or otherwise, are not to be entered into lightly. They are induced by temporarily untethering our minds from the cognitive, sensory and social inputs that in ordinary consciousness anchor them to reality. As in sleep and dreams, free from the usual waking constraints, they provide an opportunity for the brain to regain some of its former flexibility – to optimize its models of the world which over time become hidebound and inefficient.[30] But to reboot your brain and regain your sanity, you really do have to 'go a little crazy'. It's important, therefore, to ensure the 'set and setting' are appropriate: that you enter into the experience with the right attitude and plan ahead to make sure you will be in a safe place where there are caring people on hand to provide support if needed.

Ego-busting psychedelic trips and meditation retreats are not for everyone, and even those who have undergone such rites of passage may not wish to repeat them. Nonetheless, for most of us there remains an urge to keep our minds flexible as we grow older. Our brains evolved to model our environment with increasing accuracy, so with age comes experience and wisdom. But it also becomes progressively more difficult to adapt to changing circumstances, think creatively and acquire fresh skills, such as learning a new language or changing the direction of one's career. As the decades pass and our brains faithfully mould themselves to the narrow conditions in which we find ourselves in our professional and private lives, they are slowly but surely ossifying. No wonder artists, mathematicians and scientists often do their best work in their twenties.

Fans of microdosing – taking sub-perceptual doses of a psychedelic such as LSD or psilocybin every two or three days – believe it's a highly effective way to maintain their mind's youthful flexibility. They are convinced that ingesting as little as one-twentieth of the dose needed to produce hallucinogenic effects not only improves their mental well-being but also gives them an edge in terms of energy levels, productivity, analytical thinking, creativity and mindfulness. Over the past few years the practice has become popular in the intensely competitive hi-tech industries of Silicon Valley, partly inspired by tales of LSD's role in the creative epiphanies of computing giants in the late sixties and early seventies, such as Douglas Engelbart (inventor of the computer mouse, among many other user-interface innovations), Steve Jobs and Bill Gates.

Microdoses are also reputed to have sparked scientific breakthroughs, not least the discovery of the structure of DNA by Francis Crick and James Watson in 1953. The evidence for this is admittedly scant and indirect. According to a story in the *Mail on Sunday* published ten days after Crick's death in 2004, Crick told a fellow biochemist – who then told a friend – that he and other Cambridge academics took LSD microdoses in the fifties 'to liberate them from preconceptions and let their genius wander freely to new ideas'. Crick supposedly visualized the double helix of DNA while on a low dose of the drug.[31] In later life, Crick became well known for his liberal views, wild parties and fondness for cannabis and LSD, though we will probably never know whether he was an early adopter of microdosing.

Today's microdosers often cite a pilot study of psychedelics and creative problem-solving conducted in 1966, though strictly speaking this involved doses high enough to cause perceptual changes.[32] Under controlled laboratory conditions, a team led by the engineer and futurist Willis Harman at the International Foundation for Advanced Study in Menlo Park, California, gave 200 milligrams of mescaline to twenty-seven men in professions ranging from engineering, physics and mathematics to architecture and furniture design. Standard psychological tests before they took the drug and after its effects kicked in indicated enhanced pattern recognition and ability to focus. The men were then asked to spend the next four hours tackling a difficult problem they had been working on for at least three months prior to the study.

Many of the creative solutions they came up with under the influence of the drug apparently stood up under sober scrutiny. We're told that among the concrete outcomes were a design for a commercial building that was accepted by a client; space probe experiments to measure solar properties; 'a linear electron accelerator beam-steering device'; technical improvements for a magnetic tape recorder; a new conceptual model of the photon; and a mathematical theorem concerning electronic logic gates.

Participants later reported that in the quiet safety of the lab, their concentration had been enhanced, their thought processes had been faster and more flexible and they were able to see their problem in its wider context. They had a heightened capacity for mental imagery and fantasy, and were able to merge mentally with the objects and processes with which they were grappling, almost *becoming* the problem. At the end of the session they were driven home and given a sedative pill to take in case they had trouble sleeping. But in many cases, Harman wrote later, 'they preferred to stay up until well after midnight, continuing to work on insights and solutions discovered earlier in the day'.[33] Harman concluded:

> Assuming that these findings are eventually substantiated by additional research, they find their most obvious application to problem solving in industry, professional practice, and research. Here the procedure could play a role similar to that played by consultants, brainstorming, synectics [a systematic approach to problem-solving in a group of people], and other attempts to augment and 'unstick' the problem solver's unsuccessful efforts.

The psychologists were poised to conduct a follow-up study in which an active placebo would be given to half the subjects and the psychedelic to the other half in a 'double-blind' design (neither researchers nor subjects would know who took which), but it had to be abandoned after the US Food and Drug Administration (FDA) declared a moratorium on all psychedelic experimentation on healthy subjects.

The follow-up never took place. Nonetheless, one of the psychologists involved in the pilot study, James Fadiman, became an evangelical advocate of the benefits of microdosing for creativity and well-being. Another is Amanda Feilding, director of the Beckley Foundation who appeared in Chapter 5, an artist and enthusiastic microdoser in the sixties. 'I read Freud on LSD. I worked on LSD, addictively,' she told me when I interviewed her at Beckley Park. For relaxation, she and her partner would play the ancient Chinese board game Go. 'It's very intuitive and all about spatial pattern recognition,' she said. 'There's no chance involved so it's a very good test of creativity. We were passionate players. We played thousands of games and if I played on LSD and maintained my sugar levels, I won more games.'

It may come as a surprise to today's growing community of microdosers, however, that at the time of writing not a single placebo-controlled study of the practice has been published. Most of the enthusiasm to date has been based on anecdotal reports in the blogosphere and preliminary research such as Harman and Fadiman's that predates the legal clampdown. This matters because the cerebral boost that microdosers experience could be a placebo effect, enhanced by the drugs' legendary status as mind-altering substances and press coverage of research suggesting cognitive and mental health benefits from higher doses.

This striking lack of evidence is about to change, however. Feilding told me that one of her remaining ambitions was to scientifically test the cognitive effects of very low doses of LSD and, as I write this, researchers at Beckley and Imperial College London have just launched the world's first randomized, placebo-controlled study of microdosing. People who are already microdosing at home will be invited to follow instructions to set up the experiment for themselves. They will put either a microdose or nothing in opaque gel capsules, label them with QR codes, randomize them and swallow one every three days. Throughout the study they will fill out questionnaires and play cognitive games online to gauge their focus and well-being. The researchers say their innovative 'self-blinding' or 'DIY placebo' design, recruiting volunteers who are already microdosing on their own initiative using drugs they have sourced themselves, will get around the bureaucratic, costly legal restrictions that currently govern psychedelic studies.

A second, crowd-funded study will test Feilding's theory that a microdose of LSD can give someone an edge in tasks involving pattern recognition, such as playing Go. Forty-eight participants will be randomly assigned to receive either 15 micrograms of LSD twice a week for four weeks, or a placebo. Mood, cognition and brain function will be tested at the start and end of this period and on dosing days, including pitting their wits against a Go-playing AI program. The researchers will use EEG to assess whether similar changes in brain activity to those already seen with higher doses of LSD underlie any improvements in cognition.

Until data from these placebo-controlled studies has been published, the jury remains out on microdosing. The most thorough observational research to date, conducted by psychologists at Macquarie University in Sydney, Australia, and involving almost a hundred microdosers, does suggest the practice leads to significant improvements in self-reported productivity, creativity, happiness and connectedness on the day the drug is taken. These boosts in performance did not carry over into the following two days, however, as microdosers commonly believe happens, apart from slight improvements in productivity. Nonetheless, the practice did seem to have long-term mental health benefits. Compared with the start of the study, six weeks later there were reductions in depression, stress levels and mind-wandering.[34]

Tellingly, the microdosers in this study also developed an increased capacity for absorption: the tendency to have intense, deeply immersive imaginative experiences. People who have high scores on the standard test of this personality trait are easily caught up emotionally in stories, films and aesthetic experiences such as enjoying the great outdoors, watching a sunset, reading poetry or listening to music. They may get so engrossed in these activities that they become unaware of everything else, perhaps losing track of time, forgetting to eat, sleep or go to the bathroom. For them it can feel as if their ego has been subsumed into the imaginative activity or object of attention, which acquires an aura of meaningfulness that has been described as 'hyperreality'.[35]

An increased capacity for absorption may be one of the most important factors underlying the long-term benefits of both psychedelics and meditation. And following a retreat or trip, any kind of activity that entails deep absorption, such as listening to and playing music, yoga, dance, artistic pursuits and literature, may help keep the doors of the mind wide open.

Absorption has been a common thread running through all the altered states explored within these pages. As you may recall from Chapter 5, people who are naturally prone to absorption are more likely than others to have inherited a gene that makes an unusually 'sticky' version of the serotonin 2A receptor (see page 134). These individuals' 2A receptors have a high affinity for serotonin, which may contribute to their tendency to get caught up in imaginative experiences. These are the receptors that psychedelic molecules latch onto, so in theory the finding could also explain how microdosing might heighten a person's ability to develop and maintain absorptive states of concentration, through increased binding at these sites.

There are further hints of a common underlying neural theme.

As we saw, individuals who score high on the standard test of absorption – the Tellegen Absorption Scale – are unusually prone to hypnotic suggestion; and recent research, perhaps unsurprisingly, suggests that people become more suggestible under the influence of a psychedelic. Meditators who can attain the deepest contemplative states also rank higher than average on the Tellegen scale.[36,37]

Graham Jamieson, a cognitive neuroscientist who studies hypnosis at the University of New England in Australia, believes both meditation and hypnosis owe their startling effects to our species' innate capacity for absorption: our ability to temporarily screen out distractions in order to focus exclusively on our mental simulations.

I crawled out of bed very early one morning in spring 2018 to ask Jamieson to elaborate – it was 6 a.m. in the UK but early evening in Australia when I picked up the phone. Absorption, he explained, is what underlies flashes of intuition, the almost mystical, 'noetic' sense of revelation that accompanies the discovery of some deeper truth. 'Those intuitions can occur in mathematics, they can occur in science, they can occur in creative endeavours where there is literally a transformation in consciousness,' he said. 'There is an insight, an expression of a basic property of human consciousness, which is to *know*.' Jamieson isn't just talking about rare strokes of genius, but also the everyday moments of aesthetic pleasure, creativity and insight that can arise whenever ordinary people enter a flow-like state of consciousness. He said: 'It includes getting absorbed in the beauty of nature or music. It can happen when you're a student learning maths or learning computer programming or doing a physics experiment in high school. These moments of insight are moments of absorption, moments of noetic consciousness, and the same underlying mechanism makes them possible.'

In ordinary consciousness, Jamieson said, our cognitive resources are continually being hijacked by unexpected sensory stimuli such as a bump in the night, a stop sign while driving, feelings of hunger or thirst, or physical discomfort. Our lives may depend on this. There are times, however, for example at work or college, when we need to screen out unimportant sensory stimuli such as birdsong, noisy plumbing or the sound of someone putting a kettle on in another room. It's at times like these when there could be enormous potential benefits from ignoring the insistent voices of our senses clamouring to be heard. To maintain the quality of deep concentration that leads to insight, said Jamieson, the 'gain' or 'precision' of these distracting prediction errors must somehow be reduced. Their progress upwards through the predictive hierarchy and into conscious awareness must be blocked.

Exactly how this happens has yet to be determined, but Jamieson suspects that the telltale footprint of absorption in the brain may be an intense burst of activity in the anterior insula, where inferences about internal bodily stimuli enter conscious awareness. This may act like a red light temporarily preventing competing sensory messages – salient signals that under normal circumstances would demand our attention – from intruding into consciousness. He has proposed that the same mechanism of sensory traffic control underlies suggestibility in hypnosis and focused attention in meditation.[38] In the former it involves concentrating all your trust and attention on the sound of the hypnotist's voice. In the latter it involves focusing exclusively on a bodily sensation such as the breath.

We do the same, albeit unconsciously, whenever we deploy our imagination and creativity, becoming absorbed in some narrative fantasy or mental simulation, because to do these things successfully we must ignore the competing data from our senses even as they insist that we are in fact sitting in a cinema, in an armchair or at our desk in the office. In order to sleep at night, and dream, an even more far-reaching gating of sensory information is required, because only when sensory inputs have been completely disconnected can overnight maintenance – the optimization of our cognitive models – begin.

So, absorption is akin to dreaming with your eyes wide open, though with an altogether different purpose: creative insight. For it to get going you have to have a profound feeling of safety, Jamieson explains. 'You just don't get absorbed when you're about to get shot. At a very profound level you have to feel at home in your world, at home in your relationships and your life. Ultimately you have to be at home in your body.' Jamieson believes experiences like these are intrinsically valuable because they enrich our lives with a profound sense of *meaning*. 'The sorts of mental training we engage in and the way we habitually learn to direct and regulate our attention determine the extent to which we can engage those states of absorption,' he said. He's not just talking about Eastern spiritual traditions like yoga and meditation. The same mental disciplines are found in Christian and Muslim contemplative prayer, and they underpin creative reverie in the visual arts, in science, music and dance.

'Understanding these mechanisms,' said Jamieson, 'then using that understanding to help people have a greater capacity to access and develop those states of consciousness, is something that can add huge value to human lives and to society.'

In an age of almost incessant hi-tech distractions, not least from our smartphones, how wonderful it would be if more people were given the chance to learn how to enter these states of absorption or 'flow' at will. The potential benefits for our productivity and effectiveness at work – not only for engineers, mathematicians and architects, but also teachers, doctors, drivers, musicians, managers and shop floor workers – would be enormous.

More importantly, Jamieson would argue, the ability to focus our minds allows us to flourish as human beings. In 2010, the Harvard psychologists Daniel Gilbert and Matthew Killingsworth published a study suggesting that our happiness depends more on what we're thinking at any particular moment than what we're doing, and that mind-wandering is a cause of negative emotions such as momentary depression and anxiety.[39] 'A human mind is a wandering mind,' they concluded, 'and a wandering mind is an unhappy mind.' The irony is that to gather their data the psychologists used a smartphone app to interrupt what people were doing at random moments throughout the day to ask them questions such as, 'How are you feeling right now?'

If the new science of absorption teaches us anything it is that we should learn how to tune into our bodies, perhaps through mindfulness and meditation, with or without the help of psychedelics – and turn off our smartphones.

# 9

# *The Void Between Dreams*

If a survey published in 2018 is to be believed, Tibetan Buddhist monks are more afraid of dying than Indian Hindus, lay Tibetan Buddhists, American Christians and even people who profess no religious faith at all.[1] According to the survey, despite their insistence that they don't believe in an unchanging self or soul, the monastics were almost twice as scared of self-annihilation as any other religious or non-religious group. Even when participants' age was taken into account (older people might be more frightened of death due to its greater proximity) the result stood.

Buddhists hold that there is no solid, permanent self and that to realize this truth liberates one from the fear of death. The British philosopher Derek Parfit reached the same conclusion, arguing that this realization can dispel existential angst and break down barriers between people.[2] This was certainly true in his experience. In his book *Reasons and Persons* he wrote: 'My life seemed like a glass tunnel, through which I was moving faster every year, and at the end of which there was darkness.' But when he stopped believing that within him resided an indivisible, fixed self, he felt liberated. 'When I changed my view, the walls of my glass tunnel disappeared. I now live in the open air.'

So how come it isn't working for these Tibetan monks? In the survey, which involved 520 participants living in the US, India and Bhutan filling out a battery of psychological questionnaires, the vast majority of the sixty monastics who took part (95 per cent) claimed that their absence of belief in an unchanging self was helping them to cope with the prospect of dying. And yet their answers to questions about fear of death suggested otherwise.

I was taken aback when I first read these findings in the Buddhist magazine *Lion's Roar*. I had assumed that by dispelling the illusion of fixed selfhood, meditation liberated monastics from fear of annihilation. When I got my hands on the original paper reporting the study, in the journal *Cognitive Science*, I came across this line near the end: 'None of the participants we studied were long-term meditators.' Buddhist monks who don't meditate? Confused, I emailed one of the researchers, professor of philosophy Jay Garfield from Smith College in Hampton, Massachusetts, and the University of Melbourne in Australia: 'I wasn't aware that there are Tibetan Buddhist monastic traditions that don't involve intensive meditation practice. Have I misunderstood?'

'Contrary to popular belief,' Garfield replied, 'the vast majority of Tibetan monks are not professional meditators. That is a specialty reserved for a few advanced practitioners. Most monks teach, work in kitchens or in offices, or in publishing arms of monasteries, or study for their exams.'

Garfield and his fellow philosophers speculate that if the monks had been meditating all their adult lives, the practice would have dislodged their innate, bone-deep attachment to the idea that they have a self that continues unchanged until their dying day. 'Our sense of identity across the biological lifespan is resilient,' they write, 'and perhaps the thought of self-annihilation triggers fears too primitive to be easily tamed by the philosophical belief that there is no persistent self.' Hindus and Christians can console themselves with the notion that they have a soul that will either be reincarnated or continue into an afterlife. Some Christians believe that, come the Day of Judgement, even their physical bodies will be restored. But for Buddhists there is no resurrection, no soul, no persisting self. Their only consolation is that a naked stream of consciousness and karma will trickle from this life into the next.

For monks who don't practise meditation and are still unconsciously attached to the inbuilt notion of solid selfhood, write the philosophers, the doctrine of 'non-self' may worsen their fears of annihilation. If they can secure the necessary funding, they hope to test this hypothesis by including experienced Buddhist meditators in their next survey.

The science reviewed within these pages suggests that there are some harmful habits of mind so deeply imprinted on our psyches that only a profound, temporary alteration of consciousness – through meditation, extreme physiological or psychological stress, virtual reality, hypnosis or psychedelics – can erase them. No amount of positive thinking can cure an episode of devastating depression, the unquenchable thirst of alcoholism or the gnawing fear of death. Perhaps unconscious biases or stereotypes regarding things like race, class, gender and sexuality that we pick up, like mental parasites, on our journey through life are similarly immune to reason and noble intentions? Implicit Association Tests, which measure the ease with which people associate particular categories of person with particular traits, suggest a majority of Americans (both young and old) are biased against older people, for example, which might unconsciously affect their decisions in situations such as job interviews. There may be times when the mismatch between our noble aspirations and the emotional inferences carved into our brains' predictive models is too great to overcome.[3]

Easing worries about death is a poignant example of how altered states can dislodge long-established but ultimately unhelpful patterns of thinking. Psychedelics were first used to facilitate psychotherapy in the fifties, with an early focus on alcoholism. At sufficiently high doses, they can occasion mystical experiences, and these are intimately bound up with their efficacy as treatments not only for addiction and depression but also existential anxiety. We are only just beginning to rediscover their potential for easing life's final transition – a potential that Aldous Huxley was speculating about more than half a century ago. In 1958, in a letter to the psychiatrist Humphry Osmond (who coined the term 'psychedelic' and provided Huxley with mescaline for the famous trip described in *The Doors of Perception*), he suggested that LSD could be offered to terminal cancer patients to make their dying a 'more spiritual, less strictly physiological process'.[4]

In Huxley's final novel *Island*, published in 1962, he envisaged an ideal society in which life's pivotal transitions are facilitated by '*moksha* medicine', a psychedelic derived from mushrooms. *Moksha* is a Sanskrit term meaning liberation. On his fictional island of Pala, adolescents on the brink of adulthood, people undergoing a psychological crisis and the dying take the medicine as a healing sacrament. A year after the book came out, on the morning of his own death from cancer of the throat, Huxley scrawled on a piece of paper 'LSD – try it – intramuscular – 100 mcg'. His wife administered the injection, against the advice of his doctors, and he died peacefully a few hours later.[5]

In the sixties, some US clinics began using LSD, in conjunction with counselling, as a treatment for existential anxiety in patients recently diagnosed with cancer, with promising results, but their work became increasingly difficult from 1966 onwards as state-level bans on the drug came into force. Half a century would pass before the tide of public and political opinion began to turn again in favour of psychedelic-assisted therapy. The first glimmer of progress was a small study published in 2014 by Swiss and American researchers. Led by Peter Gasser at the University of Bern, it investigated LSD with psychotherapy to ease anxiety associated with life-threatening conditions including Parkinson's disease and cancer. The vivid mental imagery, powerful emotions and enhanced recall of memories – that psychiatrists in the sixties came to believe complemented talking therapy so well – only occur at doses higher than 100 micrograms, so Gasser and his team compared a dose of 200 micrograms with a microdose of 20 micrograms, which served as an 'active placebo'. They found that after two LSD-assisted therapy sessions, patients in the higher-dose group became significantly less anxious. Remarkably, the improvements were still apparent a year later.[6]

This was a pilot study involving only a dozen people, but by providing preliminary evidence that taking a psychedelic in a controlled, therapeutic environment can be safe, it paved the way for larger studies. A trial of twenty-nine people with life-threatening cancer followed in 2016, finding that a single dose of psilocybin with psychotherapy led to significant improvements in anxiety and depression that lasted at least six months.[7] Patients who took the drug became less afraid of dying and enjoyed an increased quality of life.

In common with many other studies of psychedelic therapy, there was a positive correlation between beneficial effects and the intensity of mystical experiences that people had under the influence of the drug. If larger clinical trials can confirm the safety and efficacy of this approach, perhaps one day psychedelic therapy will become a standard option for all terminally ill patients. Some doctors and psychiatrists may baulk at the idea of 'prescribing' a mystical experience, but I suspect the welfare of their patients will trump any philosophical qualms they might have, because while modern medicine has been hugely successful at prolonging life far beyond its natural span, thus far it has had little to offer for easing the psychological trauma of its ending, apart from mind-numbing tranquillizers.

In common with meditation, one of the spiritual insights psychedelics provide entails a loosening of our lifelong identification with our physical bodies – the very core of selfhood. To break the spell of bodily selfhood, in addition to spending long hours deep in meditation, monastics in the Thai Forest Tradition deploy a range of ascetic practices. Human skeletons are sometimes displayed in the temples of monasteries in Thailand to remind monks of the transitory nature of their bodies and, when a monk dies, his corpse may be borne to a room where it will lie unpreserved in the tropical heat for several days so his former companions can meditate on impermanence in its disintegrating presence.

Novices in this Buddhist tradition are taken on regular educational trips to their local morgue. In a Dhamma talk during a retreat I attended a few years ago, a middle-aged monk recalled his experiences as a novice visiting a police morgue in Bangkok where crude post-mortems were performed on the bodies of people who had died overnight, often violently, on the city's streets. This monk's unsentimental, almost gleeful description of the butchery he witnessed on these visits is testament, perhaps, to the efficacy of the training. After several visits, he told us, his initial disgust and horror had been replaced by a kind of fascinated awe as the raw, physical stuff of life was exposed. He found the vivid, iridescent colours of intestines particularly beautiful.

One or two retreatants listening to the monk's talk that evening fled the shrine room in a state of queasy distress, and the next day there were complaints of 'insensitivity'. Death, let alone dead bodies, remains profoundly taboo in Western societies, so it is hard to imagine such practices ever catching on as ways to reduce ordinary people's identification with their physical bodies and fear of death.

There may be a gentler alternative: virtual out-of-body experiences. As we saw in Chapter 3, a part of the brain called the temporoparietal junction or TPJ creates our feeling of embodiment by integrating information from our senses, and this feeling – a fundamental dimension of selfhood – is easily bent out of shape (see page 76). Psychologists have speculated that out-of-body experiences, which can occur during a brush with death and occasionally arise spontaneously, may help people overcome fear of dying through their apparent demonstration that consciousness can exist independently of the body. As many as one in five people who survive a cardiac arrest report experiences such as the sensation of looking down on their body from above, travelling down a dark tunnel towards a bright light, reviewing their lives in a flash from start to finish, and feelings of euphoria or peacefulness. People often come back from this state profoundly changed, with a more generous, compassionate outlook on life, reduced fear of dying and increased belief in an afterlife.

Is it the out-of-body dimension of the experience that reduces existential anxiety? Clinical psychologists at the University of Barcelona recently tested this hypothesis on thirty-two female volunteers, using immersive virtual reality to create a 'full body ownership' illusion. For sixteen of the participants this was followed by an induced out-of-body experience.[8]

The women were invited to recline in a comfortable chair in the lab, put their feet up and don a virtual-reality headset – whereupon they found themselves seated in a virtual living room. From a first-person perspective, they could look anywhere in the room including down at their virtual torso, arms and legs, with their feet up on a virtual coffee table, and across the room to a mirror where they could see their reflection. Real-time motion capture allowed the avatar's limbs to move in synchrony with their own. And floating in mid-air a few inches from their virtual bodies the women could see four spheres about the size of tennis balls, which occasionally bumped into them. Whenever this happened, a 'vibrotactile stimulator' instantaneously nudged the corresponding part of their real, flesh-and-bone body.

In the same way people can be persuaded to treat a rubber hand as if it were their own (see page 79), multisensory integration in the volunteers' brains created the illusion that the body in this looking-glass world was theirs, even though, on an intellectual level, they knew perfectly well that it wasn't. They began to *identify* with the computer-generated avatar, investing it with some of their selfhood. Then something changed. Their visual perspective rose several feet into the air until they were looking down on it from somewhere near the ceiling.

For sixteen of the women, the floating balls rose to the ceiling with them and they continued to feel their impacts. This broke their identification with the body below, which no longer moved in concert with their movements. These women had a full-blown out-of-body experience. For the other sixteen, who acted as controls in the experiment, the floating balls remained below and they felt the impact whenever they saw one bump into the virtual body. Also, whenever they moved their real arms and legs, they could still see the avatar moving its arms and legs in synchrony. For these participants, their attachment to the virtual body remained strong even though they were looking down on it from above.

After taking off the virtual-reality headset, all the subjects were invited to fill out a standard psychological questionnaire designed to quantify fear of death. Sure enough, in line with the scientists' hypothesis that an out-of-body experience would ease existential anxiety by providing implicit evidence that consciousness can exist outside the body, the sixteen subjects who had just had one reported that they were much less scared about dying than the sixteen who had not.

So it seems even a simulated out-of-body experience can reduce anxieties about our inevitable demise. Near-death and out-of-body experiences challenge our brains' unconscious predictive models of bodily selfhood – models that began to take shape all those years earlier when we kicked out for the first time in our mothers' womb. They are relearning experiences: opportunities to update these deep inferential models to allow for the possibility of existence without a body. Regardless of whether or not there is life after death, out-of-body experiences implicitly teach us to *expect* it.

The Dalai Lama has said that one of the insights afforded by a lifetime of contemplative practice is that after death will come rebirth.[9] Buddhist monastics' unshakeable conviction that their consciousness will survive the dissolution of their bodies and be reborn into another physical form – animal, human or divine – may stem from experiences of 'non-duality' during meditation, when the distinctions between their physical body and its surroundings, the self and everything else, dissolve. In the immaterial jhana states outlined in the previous chapter, conscious awareness remains even after all sensation, thought and emotion have ceased (see page 216). Could there be a better demonstration, this side of the grave, of the possibility of continued existence after death?

Deep sleep – the interval between our dreams – is a strong candidate. For the vast majority of people this is an unconscious state. When researchers wake up volunteers in a sleep lab during non-REM sleep and ask them what they remember from a moment earlier, they nearly always draw a blank. If, however, you were to awaken a sleeping Tibetan Buddhist monk practising a meditative technique called Bardo yoga – in which consciousness is maintained during non-REM sleep – and ask him the same question, he would probably answer '*'od gsal gyi sems*' – 'the clear light mind'.

This is conceived as perfectly enlightened, pure awareness, an existence beyond space and time little different from the intermediate state between consecutive lives. Every dimension of human selfhood – bodily, social and autobiographical – has vanished. The practice of sustaining pure awareness in the void between dreams originated in pre-Hindu Ancient India. As the Sanskrit scriptures known as *The Upanishads* describe it:

> Then a father is not a father, a mother not a mother, the worlds not worlds, the gods not gods, the Vedas not Vedas. Then a thief is not a thief, a murderer not a murderer… He is not followed by good, not followed by evil, for he has then overcome all the sorrows of the heart.[10]

In Bardo yoga, non-REM sleep is a training ground for death, the idea being that the meditator retains full awareness in order to familiarize him or herself with the limbo between this life and the next. Tibetan Buddhism teaches that people who haven't done this can become confused and disoriented in the dimensionless consciousness that follows death, whereas someone who has practised Bardo will retain the composure needed to ensure a favourable rebirth into a higher plane of existence.

With effort and dedication, in theory anyone can practise Bardo yoga. One way to access this realm beyond personal identity and the physical world, according to sleep yoga teacher Andrew Holecek, is to close your 'eyes' during a lucid dream, carrying this lucidity with you into dreamless sleep.[11] One of the most effective ways to promote lucid dreaming is to take up meditation.[12] Until recently, however, most neuroscientists would have scoffed at the idea that anyone could retain sufficient awareness during non-REM (mostly dreamless) sleep to pursue any conscious goal. If they'd even heard of it, they probably would have dismissed Bardo yoga as mystical mumbo jumbo.

But a study published in 2013 provided strong, albeit indirect, evidence that meditators who have clocked up thousands of hours of practice over the course of their lifetime retain a degree of consciousness during this phase of the sleep cycle. Neuroscientists at the University of Wisconsin-Madison recorded significantly greater electroencephalography (EEG) activity in the gamma frequency band (25–40 Hz) in the brains of experienced meditators during non-REM sleep compared with people who had never meditated. The longer they had been practising daily, the higher the amplitude of gamma in their occipital and parietal lobes (responsible for sensory processing and integration, respectively).[13] Several years earlier the same lab discovered intense gamma activity during waking meditation in Tibetan Buddhists who had been practising regularly and intensively for more than fifteen years. Even when they were resting in between meditation sessions, the amplitude of gamma waves coursing through their brains was as much as twenty-five times greater than in controls.[14]

Gamma waves are associated with consciousness, in particular 'secondary consciousness', characterized by self-awareness and metacognition. You may recall from Chapter 2 that people who are having a lucid dream have increased gamma activity in their brains, and that applying weak electrical pulses at this frequency to the scalps of experimental subjects during REM sleep induces lucid dreaming (see page 51). In waking consciousness, gamma is implicated in attention, learning and memory – the ascending flow of sensory prediction errors that informs perception and updates our cognitive models of reality. While alpha waves appear to orchestrate the brain's top-down predictions and active inferences about sensory inputs, gamma waves handle information flowing the other way. They synchronize the activity of far-flung sensory processing regions to provide 'Aha!' moments of recognition, bundling different streams of data (for example the colour, brightness, shape and orientation of an object) to create a single conscious impression.

Gamma oscillations are also involved in *imagining* such impressions, so it's easy to see how the same high-frequency waves might summon the visions we see in our dreams. But whereas in ordinary folk like me or you, bursts of gamma in waking consciousness and dreams only last for fractions of a second, in the yogis studied at the University of Wisconsin-Madison they continued to reverberate across their brains for minutes on end. This highly unusual electrical activity may correspond to the heightened state of mindful awareness or 'spaciousness' that adept meditators often describe: the way it must feel when one's senses are wide open to the richness of experience as it unfolds, uncontaminated by judgement or the desire to change anything (the stuff of active inference).

Yogis may maintain some of this spacious awareness – as revealed by the strong gamma waves in their sleeping brains – even in non-REM sleep after the gates of their senses have been securely shut. In the light of this new research, the most sceptical neuroscientist would have to admit the possibility that Tibetan monks can maintain sufficient awareness in non-REM sleep to practise Bardo yoga: preparation – as they see it – for the intermediate state between this life and the next.

For these yogis it must seem as though they are having a near-death experience every night in their dreams, and among the benefits may be less fear of death and increased belief in rebirth. In fact, the 'clear light mind' of Bardo and the burst of ecstatic, heightened consciousness reported by people who have returned from the brink of death are starting to look uncannily similar.

Near-death experiences are not a purely Western obsession fuelled by airport bestsellers and Netflix documentaries. They have been reported across cultures and are universally described as highly vivid and lucid – 'realer than real'. Research into the neural correlates of the phenomenon in humans has been thin on the ground, but in 2013 a team of doctors and scientists at Johns Hopkins School of Medicine in Baltimore, US, discovered what may be the equivalent in rats.[15] Using EEG, they detected a surge in gamma waves in the animals' brains seconds after an induced cardiac arrest and shortly before death. Even though the animals' hearts had stopped pumping oxygenated blood around their bodies, the amplitude of gamma waves in their brains was briefly greater than in the normal waking state.

'The data suggest that the mammalian brain has the potential for high levels of internal information processing during clinical death,' the researchers reported in the prestigious journal *Proceedings of the National Academy of Sciences*. 'The neural correlates of conscious brain activity identified in this investigation strongly parallel characteristics of human conscious information processing.'

Anecdotes abound of brief EEG spikes on hospital monitors as people die in intensive care, but the only formal investigation of the phenomenon to date was conducted at George Washington University Medical Center in Washington DC around ten years ago. It involved seven terminally ill patients, five women and two men, who had their life support withdrawn on compassionate grounds.[16] Seconds after their blood pressure had fallen to undetectable levels, monitors showed a surge in overall electrical activity in the brains of all seven people that lasted up to three minutes. During the spikes the patients were motionless and had no pulse, yet the electrical activity suggested they were experiencing a brief interlude of conscious awareness before death, in spite of heavy sedation. All the patients were declared dead shortly afterwards. In a report published in 2009, the doctors describe how they managed to retrieve raw EEG data from the device used to monitor the brain activity of one of the patients (the others were wired up to another kind of device), which showed a burst of high-frequency gamma waves 'consistent with cerebral arousal'.

When people fortunate enough to survive a near-death experience describe what happened, their accounts resemble a psychedelic trip – including time distortion, reviewing autobiographical events ('my entire life flashed before my eyes'), sudden insights, disembodiment, spiritual feelings of otherworldliness, 'unity' (being at one with everything) and encounters with significant others or supernatural beings. Recent studies also reveal intriguing overlaps in the long-term positive effects of psychedelics and near-death experiences, including increased connectedness with nature, improved psychological well-being – and less fear of dying. These shared features may be no coincidence. Psychedelic molecules trigger their extraordinary acute effects by binding to serotonin 2A receptors in the cortex, and it seems likely that a massive increase in binding at these receptor sites also precipitates near-death experiences.

Rick Strassman, a psychiatrist at the University of New Mexico School of Medicine, has proposed that during birth, in our dreams and when we're close to death the pineal gland near the centre of the brain releases a flood of DMT, the hallucinogenic component of the South American brew ayahuasca and the brain's own home-grown psychedelic.[17] It's an attractive idea but almost certainly false. A study published by researchers at Imperial College London in 2018 did find many similarities between near-death experiences and the rapid, intense transformations in consciousness that occur in the minutes after being injected with DMT in the lab, including ego dissolution, disembodiment, sensations of joy and peace, visiting other worlds and encountering mystical beings.[18] It's also true that trace amounts of internally synthesized, 'endogenous' DMT have been detected in cerebrospinal fluid and the pineal gland. But the gland would need to churn out much more, about 25 milligrams of DMT, within a couple of minutes of a cardiac arrest to generate a powerful psychoactive effect, according to the pharmacologist and renowned psychedelic researcher David Nichols from the University of North Carolina, Chapel Hill.[19] The principal role of the pineal gland, which is the size of a pea and weighs only 100–180 milligrams, is to produce 30 micrograms (millionths of a gram) of the sleep hormone melatonin in the course of a day – roughly a thousandth of the mass of DMT it would need to produce within minutes of a person's heart stopping.

'Although the romantic notion that DMT is released from the pineal gland to produce altered states of consciousness at various times of stress is appealing to some, science and logic suggest that other, more well studied systems provide more sound explanations for out-of-body experiences,' Nichols told psychonauts at the Breaking Convention psychedelics science conference in London that I attended in 2017.[20] Strassman has suggested that DMT could be manufactured in advance elsewhere in the body (for example in lung tissue, which contains the necessary enzyme) and transported to the brain where it would be stored ready for rapid release in states of extreme physiological stress. But Nichols pointed out that when injected or smoked, the molecule is broken down within minutes by the body's monoamine oxidase enzymes (whose normal function is to regulate levels of serotonin and other monoamine neurotransmitters). The same fate would befall endogenous DMT wherever it was produced. What's more, he said, there is no plausible evidence that the molecule can cross the blood–brain barrier or that it is sequestered in the fluid-filled sacs or 'vesicles' of synapses, as has been proposed by advocates of the near-death DMT theory.

His explanation for the surge in gamma waves and conscious awareness in the minutes after cardiac arrest or asphyxiation doesn't involve a sudden release of endogenous DMT – which he doubts plays any role in human physiology – but rather a brainstorm of regular neurotransmitters whose function is well known:

> Something up here knows that you have a serious problem down here. You hear about people being in serious accidents when time slows down and they watch things happen very slowly and they have time to interact and act. I think this activation of cortical function is related to a survival mechanism where the body knows it's about to die and the brain becomes very active, pulls out all the stops, starts running on all eight cylinders and says 'What can we do to save ourselves?'

Levels of the neurotransmitters noradrenaline, glutamate and dopamine are known to increase dramatically in the seconds following cardiac arrest, said Nichols, which would promote arousal and alertness. Add to this heady cocktail a dash of endorphins, the body's own painkillers responsible for the 'runner's high', and the result could be euphoria and an out-of-body experience. But what of the profound spiritual insights and encounters with deceased relatives or supernatural beings? What of the lasting changes in personality and outlook that often occur in people who have escaped from the jaws of death, such as increased concern for others, greater appreciation of nature and reduced interest in social status and possessions? Research suggests these changes are significantly more likely to occur in people who report a near-death experience compared with those who come close to dying but without having such an experience.[21] Surely only a powerful psychedelic such as DMT could create such otherworldly special effects and long-lasting psychological transformation?

According to David Nutt and Robin Carhart-Harris at Imperial College London, we need look no further than the neurotransmitter serotonin to explain the phenomenon of near-death experiences. In rats, serotonin surges to twenty times its normal level within two minutes of asphyxia, and a similarly rapid increase in a human close to death would ratchet up activation of serotonin 2A receptors in the brain, replicating the classic effects of a psychedelic trip. Nutt and Carhart-Harris's serotonin stress theory, which I introduced in Chapter 6 (page 162), proposes that mammals have evolved two complementary strategies for responding to challenging circumstances.[22] The first is mediated by profuse, high-affinity serotonin 1A receptors and occurs when someone is exposed to mild levels of adversity. It promotes acceptance and resilience. The second is mediated by the more sparse, less sticky 2A receptors and only kicks in when we are faced with extreme, life-threatening situations such as starvation, asphyxia or a lethal adversary. This second stress response promotes brain plasticity: the kind of radical, adaptive thinking that might ensure our long-term survival – assuming, of course, we live to fight another day.

In the game of life, fine-tuning the balance between stability and plasticity to match circumstances is the key to success. Over the past decade, as we have learned more about the neuroscience of consciousness, a picture has started to emerge of how altered states – from dreams and trance to psychedelics and meditation – shift the balance in favour of change, shattering the dominance of established, conservative models of thought and behaviour: models that in their most intractable forms are responsible for conditions such as addiction, depression, anxiety and post-traumatic stress disorder (PTSD). In the process, altered states make psychological growth and adaptation possible. We've seen how, for example, drugs such as LSD and psilocybin temporarily stifle alpha waves in the default mode network (DMN), dissolving the ego and promoting communication between brain regions and networks that aren't usually on talking terms, bringing alternative cognitive models to the fore. By contrast, years of committed meditation practice seem to work from the bottom-up, expanding consciousness by boosting gamma waves, opening up our senses and blurring the boundaries of bodily selfhood. The final result looks very similar: a brain less dominated by tired, automatic ways of thinking.

In light of these recent advances in the scientific understanding of altered states – and consciousness itself – the future possibilities for medicine and society look exciting. We have an opportunity unprecedented in human history to harness altered states as part of safe, effective new treatments for mental illness and addiction, to enhance general well-being and ease our passing. Different technologies of consciousness will almost certainly be combined in creative, synergistic new ways. It has been suggested, for example, that psychedelic drugs, which temporarily increase suggestibility, could be used in conjunction with hypnosis to channel and enhance the therapeutic effects of each.[23]

In the previous chapter, I discussed the promise of meditation and microdosing as ways to carry the benefits of a psychedelic trip forward into everyday, ordinary consciousness. The relationship between psychedelics and meditation goes back a long way, perhaps several thousand years to the soma rituals practised in Ancient India. Meditation teacher, author and former Buddhist monk Jack Kornfield has claimed that, in the sixties and seventies, psychedelics were responsible for inspiring most of today's Western Buddhist teachers to take up their spiritual practice.[24] One of the best-known psychedelic converts to Eastern spiritual philosophy is Ram Dass, formerly Richard Alpert, the clinical psychologist who worked with Timothy Leary at Harvard before they were expelled in 1963. In *The Psychedelic Experience*, Alpert, Leary and Ralph Metzner famously repurposed *The Tibetan Book of the Dead* to serve as a guide for trippers seeking enlightenment, convinced that a sufficiently heroic dose of LSD provided a shortcut to the 'clear light' – transcending selfhood, space and time – usually only experienced through the dedicated practice of Bardo yoga.

Amanda Feilding, the director of the Beckley Foundation, which collaborates with psychedelic researchers around the world, told me she would love to set up a scientific investigation of the effects on the brain of meditating under the influence of psychedelics:

> I long to do that research and find out what the crossovers and similarities and differences are. Long ago I met some very high-level meditation teachers, one very accomplished Indian meditator in particular who was a brain surgeon as well as a teacher, and he asked if he could have some LSD – this was the sixties before it was illegal – and so I gave him some. [*She laughs.*] He came back afterwards and said, 'Yes, that was very very nice, just what I know. Could I have some more please?'

> I think the drug gets you to a particular state of mind more reliably. And if you've spent the past fifty years meditating to a high level you *know* the state. You're absolutely familiar with the landscape.

Feilding would have no difficulty recruiting subjects for such a study. In the US, several Buddhist organizations advocate psychedelic use, with careful attention to set and setting, as a means to open the mind during meditation. A few are even providing ayahuasca and magic mushrooms, almost holy sacraments, during meditation retreats.[25] In addition, many people meditate without subscribing to any particular belief system so are completely unfettered by religious precepts against intoxicating substances.

Kornfield is one of the Western Buddhist teachers open to the idea of the careful use of psychedelics to enhance meditation. He believes that both reveal the startling truth that consciousness 'creates the world' – that what we consider to be physical reality is in fact shaped by our minds and not the other way round. Now that psychedelics are regaining some of their former respectability, many Western Buddhists are turning to them once again for spiritual inspiration. Some claim that the drugs have reinvigorated their practice.

Other Buddhist teachers, however, are uncomfortable with these developments. They caution that chemically-induced experiences of non-duality can only take one so far. The Thai Forest monk Ajahn Sucitto, former abbot of Chithurst Buddhist Monastery in West Sussex, UK, has written on his blog about his own adventures in psychedelic consciousness as a young man, but he warns against the temptation to use mind-altering drugs to *reject* everyday reality rather than penetrating it through mindfulness and ethical behaviour.[26] Without the Noble Eightfold Path – the framework laid down by the Buddha for grounding an individual in the highest standards of ethical behaviour, wisdom and the discipline necessary to maintain mindful awareness in ordinary life – he believes these experiences are exciting but worthless distractions. '[The Buddha's teaching or Dhamma] is a precious pearl that should not, and cannot, be handed over to another person, let alone to a drug that can only absorb you into perceptions and feelings.'

Nonetheless, to challenge a cognitive model as deep-seated as selfhood takes more than philosophical teachings. It takes a shock of some sort. The Tibetan monks who opened this chapter – who were steeped in the Dhamma but didn't meditate and were more afraid of dying than Hindus, Christians or non-believers – are testament to this. It also goes without saying that meditation is not everyone's cup of tea, let alone the ascetic disciplines of a Thai Forest monk. For many people, a psychedelically-induced alteration of consciousness may be just what is needed to create a self-transcendent moment of realization that sets you on the path towards greater well-being and happiness, including the insight that death, even though it is the end of the self, is not the end of the world.

In his poem 'The Marriage of Heaven and Hell', the eighteenth-century printer and mystic visionary William Blake wrote:

> But first the notion that man has a body distinct from his soul is to be expunged; this I shall do by printing in the infernal method, by corrosives, which in Hell are salutary and medicinal, melting apparent surfaces away, and displaying the infinite which was hid.
>
> If the doors of perception were cleansed every thing would appear to man as it is, infinite.

> For man has closed himself up, till he sees all things thro' narrow chinks of his cavern.

Medicine, science and spirituality have found a common cause, though they speak different languages. Huxley wrote in *The Doors of Perception* about the reducing valve of ordinary consciousness; neuroscientists report how alpha waves limit connectivity within the brain and make selfhood our default mode. Blake wrote about the narrow chinks we have been reduced to peering through; scientists theorize about the grip that deep predictive models of selfhood hold over our every thought, perception and action. The scientific research reviewed within these pages suggests that, used wisely, altered states of consciousness can change these models for the better. By chipping away at the chinks through which we gaze out at the world, they can let infinity in.

# *Epilogue*

*I sing the praise of dreams. Daily will I give thanks to the Highest for the freeing of the spirit of man from the labour and sorrows that are his by day. For dreams, the delight of the world, I will give praise.*
Mary Arnold-Forster, *Studies in Dreams*

The scientific evidence presented in this book holds out the promise that, within a decade, altered states of consciousness will be used not only for treating intractable mental illnesses and addictions but also for helping all of us find more meaning and purpose in our lives. In an age of disillusionment with mainstream religions, these special states look likely to play a vital role in fulfilling the deep human need to *believe*.

You may recall that in the opening chapter I wrote, 'What other animal seeks out meaning?' It was meant to be a rhetorical question, but according to the neuroscientist Karl Friston, all living creatures – from badgers, pigeons and crocodiles to plants, bacteria and mushrooms – devote their lives to the pursuit of meaning. As Friston sees it, every organism on earth is striving to reduce uncertainty in its interactions with the environment, in much the same way scientists try to tease recurring patterns from the statistical noise of their experimental data. This is what life has evolved to do: what it means to be alive.

Humans are simply at the far end of this biological spectrum. While other animals scoured their sensory data for clues about matters such as how to keep warm and where their next meal or mate was likely to come from, our earliest human ancestors began to bend their attention to increasingly long-range survival puzzles. How could they know the best time to sow their crops? Why did some people fall ill while others remained healthy? Why were they here? Was there life after death?

Friston believes that ever since life condensed from the primordial soup some 4 billion years ago it has been on a continual quest for meaning. In 2012 he wrote a paper proposing that living things endure behind the barricades of their membranes and cell walls – resisting the tendency of all matter to disintegrate into a disordered jumble of molecules – by minimizing 'free energy'. This term from information theory essentially quantifies the prediction errors or 'surprise' they experience in their dealings with the outside world: the discrepancy between what they expected and what their senses report back to them. Friston wrote that to minimize free energy, living things deploy a statistical tool familiar to any modern scientist known as 'Bayesian inference'. The mathematical details are unfortunately beyond the scope of this book (and this author), but suffice to say Friston's paper aspired to explain the origin of life and the survival of every plant, animal, fungus and bacterium in half a dozen neat equations.[1]

If Friston is correct, the search for meaning is the meaning of life. But there's a sting in the tail for long-lived creatures like ourselves whose brains encode highly complex, far-reaching models of our world and that of other people: over time as we refine them, our models get increasingly hidebound and inflexible. Like an old pair of slippers, they may be comfortable but they have lost their elasticity and are in danger of falling apart. 'As we get older and wiser we lose plasticity,' Friston told me. 'We are now wise and fit for purpose, but only for our environment. So we can't do new environments.'

This is a limitation he came up against himself in his late twenties. His early career had been in psychiatry but in the nineties he switched to neuroscience in order to study the brains of people with schizophrenia. This led to his exploring innovative techniques for crunching through the avalanche of data from newly developed brain-imaging technologies such as MRI. He quickly discovered, however, that he didn't have the necessary grounding in advanced mathematics. He had to teach himself. 'It's much easier to do the maths when you're eighteen than when you're twenty-eight,' he mused. 'But I have to confess, all the maths I use in my work I get from Wikipedia.'

I'm still not sure whether he was pulling my leg. What is certain is that the statistical techniques Friston has created for interpreting brain-imaging data have revolutionized neuroscience. By 2016 – having also become renowned for his Free Energy Principle – he was estimated to be the most widely cited neuroscientist in the world and, at the time of writing, he is a leading contender for the Nobel Prize in Physiology or Medicine.[2]

To someone working in his field, Friston's free-energy equations and the theory of life they encapsulate are beautiful. Contrary to popular perception, scientists and mathematicians value their emotional response to new ideas very highly, and what they find most beguiling in a fellow scientist's work are its simplicity, elegance and applicability in a wide range of circumstances (think $E = mc^2$). This aesthetic appreciation of theories, hypotheses and equations isn't sentimental but an innate expression of a famous rule of thumb known as Occam's razor that featured prominently in Chapter 2. The principle states that, other things being equal, an idea that works in a variety of situations and draws upon the fewest possible assumptions to explain the available evidence is more likely to be true than a narrowly focused, overly complex one. It's not enough to explain things; your explanation must also be beautiful.

Friston and Allan Hobson (the 'dream pope') believe that Occam's razor is built into the physiology of the human brain. As we saw, their proposal is that every night in our sleep the razor optimizes our models of reality by pruning redundant synapses that have sprouted during the day's learning experiences. Crucially, this streamlining prevents the models from becoming hidebound – only meaningful under strictly limited circumstances because, like an overcomplicated, clunky scientific hypothesis, they specify too many parameters. Pruning all those superfluous synapses helps maintain the brain's plasticity.

But there's a limit to how much flexibility a good night's sleep can restore. Like Friston struggling to learn advanced maths in his late twenties, as we get older it becomes more difficult to adapt to changing circumstances. We're less good at coming up with creative ways to minimize life's uncertainties. More seriously, whenever the brain develops rigid, unhelpful models of reality – like a scientist clinging to an obsolete theory in the teeth of contradictory evidence – the result can be mental illnesses such as post-traumatic stress disorder (PTSD), anxiety, depression and addiction. This book has presented evidence that, used wisely, altered states including meditation, hypnosis and even highly absorbing activities such as gaming can – in common with sleep and dreaming – wield Occam's razor to restore health, balance and plasticity to the human mind.

Psychedelics are turning out to be the sharpest altered state in this toolbox. They could revolutionize psychiatry at a time when the developed world is facing an unprecedented mental health crisis. In 2014, more people in the US died from drug overdoses than any previous year on record. In the same year, about 8 million American adults experienced symptoms of PTSD, 40 million suffered from an anxiety disorder and 15.7 million had gone through at least one major depressive episode in the previous 12 months. Globally, according to the latest figures from the World Health Organization, the number of people living with depression increased by almost a fifth between 2005 and 2015 to 322 million, making it the leading cause of disability worldwide.[3]

For the past thirty years, psychiatric medicine has been dominated by anti-anxiety drugs known as benzodiazepines and the SSRI (selective serotonin reuptake inhibitor) antidepressants. Both classes of drug work by dulling our emotional responses, leading to a less dramatic, less meaningful experience of the world. Psychiatrists often combine these pharmaceuticals with approaches such as cognitive behavioural therapy (CBT) and mindfulness-based therapies, which aim to change the unhealthy underlying mindsets behind many mental illnesses. But the drugs themselves don't work for everyone and for those patients who do benefit they are essentially sticking plasters. They don't address the 'jammed' cognitive models behind PTSD, depression and anxiety. As a result they have to be taken daily, perhaps for years on end, and they cause side effects such as weight gain, insomnia, lethargy and loss of creativity. While the drugs allow many patients to function reasonably normally, it's at the cost of losing some of their former *joie de vivre*.

By contrast, psychedelic-assisted psychotherapy involves just one or two sessions in which perception, emotion and the sheer meaningfulness of experience are amplified rather than suppressed. This is a cathartic process that compels patients to face their demons, or 'shadow side', as a South American ayahuasquero would put it. As we saw in Chapter 5, when psychedelics bind to serotonin 2A receptors they open a window of opportunity for change by disrupting normal patterns of brain connectivity. By breaking the dominance of deeply ingrained, inflexible cognitive models, this allows alternative ways of thinking and behaving to come to the fore.

Writing in a special issue of the journal *Neuropharmacology* in November 2018, the psychiatrist Jack Henningfield from Johns Hopkins University School of Medicine, Baltimore, and the pharmacist Sean Belouin of the United States Public Health Service predict that if psychedelics live up to their promise as treatments for anxiety, depression, post-traumatic stress and addiction, they will be 'one of the most momentous breakthroughs in psychiatry and medication development in decades'. After two lost generations during which research involving humans was prohibited, Henningfield and Belouin call for the drugs to be rescheduled so that they can be put through their paces in large, rigorous clinical trials. They write that if these studies confirm the results of recent preliminary trials, and research conducted in the fifties and sixties, psychedelics will transform psychiatry.[4]

On a psychological level, the efficacy of drugs such as psilocybin, DMT and LSD rests on their ability to provoke spiritually meaningful experiences, such as feelings of unity and sacredness. These experiences may provide patients with the crucial insights into their condition they need to 'unstick' themselves and move forward.[5] A pilot study published in 2015 of psilocybin to help people quit smoking, for example, found that reductions in nicotine craving were not dependent on the intensity of the drug's effects per se, but rather on the sense of meaning and spiritual significance people later reported feeling while under its influence. After just two or three psilocybin sessions, in combination with CBT, twelve out of fifteen smokers (80 per cent) remained tobacco-free six months later, easily outstripping the success rates of standard programmes for helping people kick the habit.[6]

The importance of meaning for human well-being and flourishing shouldn't come as a surprise. The more we have learned about the brain in the past decade, the more apparent it has become that our cognitive models exert a powerful psychological and biological effect. As I hope this book has demonstrated, top-down expectations of meaningfulness help determine our physiological responses and how we perceive the world in both ordinary and altered states of consciousness, as evidenced by the placebo effect, hallucinations and perceptual illusions.

In an echo of Karl Friston's proposals about the origin of life and how all creatures minimize surprise in their interactions with the environment in order to avoid dissolution, the psychologist Michael Steger from Colorado State University says the quest for meaning is 'a matter of life and death'.[7] Steger researches people's ability to find meaning and the benefits it brings them. His work and that of his colleagues reveals that, regardless of variables such as income, disability and personality, having a sense of meaning and purpose in your life – whether it comes through work, hobbies, creative endeavours, personal relationships or helping other people – is associated with increased kindness, life satisfaction, vitality and longevity. Conversely, a lack of meaning is linked to depression and a greater risk of suicide.

Altered states – not only the kinds precipitated on purpose through meditation, drumming, dance, religious trance and ingesting naturally occurring psychedelics, but also the accidental revelations sparked by brushes with death, out-of-body experiences, dreams, fevers and epileptic seizures – have been enhancing the meaningfulness of human lives for millennia. Ego dissolution during these experiences may lead to the realization that our models of selfhood, which put us at the centre of everything, are not the be-all and end-all of existence we had assumed them to be. A greater sense of connectedness with nature and other people often follows.

These insights may have an overpowering mystical or religious tone. In light of the evidence within these pages, rather than damning all such revelations as dangerous, deluded and unscientific – as many other atheists would – I prefer to side with the founder of modern psychology, William James, whose thinking about mystical states of consciousness was strongly influenced by his own experiments inhaling nitrous oxide ('laughing gas'). In 1902, he concluded in his most famous book, *The Varieties of Religious Experience*, that we should judge transformative experiences or 'mind-cures' of this kind not by their roots but by their fruits:

> Science gives to all of us telegraphy, electric lighting, and diagnosis, and succeeds in preventing and curing a certain amount of disease. Religion in the shape of mind-cure gives to some of us serenity, moral poise, and happiness, and prevents certain forms of disease as well as science does, or even better in a certain class of persons. Evidently, then, the science and the religion are both of them genuine keys for unlocking the world's treasure-house to him who can use either of them practically.[8]

In exactly the same way a good doctor or surgeon does all she can to maximize the placebo effect associated with her medicines and surgical procedures, the important thing about altered states is not whether the revelations they bring are a totally accurate reflection of reality, but whether they improve our well-being. At a time when we are facing a global mental health pandemic, the urgent question then becomes how to exploit the unique ability of altered states to restore plasticity and help people make their lives more meaningful.

Altered states are likely to be just as important in prevention as in treatment. Exploratory studies suggest that psychedelic use in healthy people, for example, is associated with greater mindfulness, openness and connectedness to other people – measures of 'positive psychology' that are known to bolster us against developing mental illnesses.[9]

In the coming years, researchers will be keen to identify factors that increase the benefits and minimize the risks when people alter their consciousness. The work has already begun. When I signed up for the 'experience weekend' in the Netherlands organized by the UK's Psychedelic Society, described in Chapter 7, I also volunteered to take part in an Imperial College London study investigating these variables. Around 650 of us who planned to participate in retreats or ingest a psychedelic for recreational purposes agreed to complete five online surveys, answering questions about our state of mind at fixed time points from one week before to four weeks after the experience. We also reported our intentions, the dosage, the setting and our mindset shortly before taking the drug.

The results of this study – the first of its kind – were published in November 2018. They suggest that, on average, our sense of well-being was improved for at least four weeks following the trip (or trips in the case of retreats lasting several days). Being comfortable with the setting, including other people present, and having spiritual or therapeutic motivations for taking part were important determinants of increased well-being. Crucially, the intensity of any mystical or peak experiences – such as feelings of disorientation in space and time, awe, humility and a sense of oneness with the universe – was also associated with greater subsequent increases in well-being, whereas the strength of imagery and hallucinations while on the drug was not.[10]

Summing up their research, the Imperial College scientists recommend setting intentions related to establishing a 'spiritual connection' and a comfortable physical and social environment in which to take the drugs. They also advise 'efforts to maximize the likelihood of mystical-type experiences'.

What exactly do they mean? They don't go into specifics, but looking ahead to a future in which psychedelics have been legalized for use under strictly regulated circumstances – following successful clinical trials – one can envisage a new type of clinic that fully recognizes the importance of spiritually meaningful, mystical experiences for promoting human well-being. Walking into an 'altered states healing centre' for the first time will feel more like entering a sacred space – a chapel or a maloca, perhaps – rather than a clinic, and the psychedelic-assisted psychotherapy on offer will be more akin to a ceremony than a doctor's appointment. There may be religious symbols and imagery on display appropriate to the faith of patients, beautiful music, candles, incense. Providing a safe 'container' for these extraordinary experiences will be paramount, ensuring patients are kept under professional supervision during their trip and that they get home safely afterwards.

Under the same roof as the psychedelic chapels and shrine rooms in these cathedrals of consciousness, professional services will be provided in the ensuing weeks and months to help people integrate their experience into ordinary life: healing circles, discussion groups, dance workshops, music and art therapy. Down the corridor there may be a meditation hall for yoga and sitting meditation practice, a therapeutic games arcade, perhaps even a sleep lab where people will learn how to become fully aware in their dreams.

All this may seem a little fanciful, but public health experts in Canada have been working for several years on proposals for safe, controlled spaces where psychedelic-assisted therapies could be provided following the drugs' decriminalization. In 2016, Mark Haden and his colleagues at the University of British Columbia and the Health Officers Council of British Columbia provided a blueprint for regulating and managing such a service. They propose setting up a Psychoactive Substance Commission to oversee the licensing, production, distribution and taxation of the drugs, and a College of Psychedelic Supervisors, which would operate in a similar way to existing medical colleges to administer professional qualifications, enforce high standards of practice and license venues. Revenue generated from sales, fees and taxes would be used to finance the regulatory structure and fund scientific research into psychoactive substances and public health programmes.[11]

After half a century of stigma, psychedelics are about to get a second chance, and not a moment too soon. At a time when humanity faces unprecedented challenges, from existential environmental threats to the global crisis in mental health, the full range of techniques for rebooting consciousness will be needed if we're all to rediscover more meaningful, less self-destructive ways of being.

# *Sources*

## *Introduction*

1. Soler, J. et al (2016). Exploring the therapeutic potential of ayahuasca: acute intake increases mindfulness-related capacities. *Psychopharmacology*, 233(5): 823–829.

2. Dominguez-Clavé, E. et al (2016). Ayahuasca: pharmacology, neuroscience and therapeutic potential. *Brain Research Bulletin*, 126(1): 89–101.

3. Thomas, G. et al (2013). Ayahuasca-assisted therapy for addiction: results from a preliminary observational study in Canada. *Current Drug Abuse Reviews*, 6(1): 30–42.

4. Sinclair, E. (2017). Can people really die from drinking ayahuasca, as announced in the media? Chacruna.net, http://bit.ly/2ApEACS, accessed 7 March 2019.

5. dos Santos, R. G., Bouso, J. C. and Hallak, J. E. C. (2017). Ayahuasca, dimethyltryptamine, and psychosis: a systematic review of human studies. *Therapeutic Advances in Psychopharmacology*, 7(4): 141–157.

6. Kejser Starzer, M. S., Nordentoft, M. and Hjorthøj, C. (2017). Rates and predictors of conversion to schizophrenia or bipolar disorder following substance-induced psychosis. *The American Journal of Psychiatry*, https://doi.org/10.1176/appi.ajp.2017.17020223, published online 28 November 2017, accessed 7 March 2019.

7. Hobson, J. A., Hong, C. C. and Friston, K. J. (2014). Virtual

reality and consciousness inference in dreaming. *Frontiers in Psychology*, https://doi.org/10.3389/fpsyg.2014.01133, published online 9 October 2014, accessed 7 March 2019.

8. *The Matrix*. Directed by Andy Wachowski and Lana Wachowski. Warner Bros, 1999.

9. Soler, J. et al (2016). Exploring the therapeutic potential of ayahuasca: acute intake increases mindfulness-related capacities. *Psychopharmacology*, 233(5): 823–829.

## 1: *Magical Thinking*

1. Ross, S. et al (2016). Rapid and sustained symptom reduction following psilocybin treatment for anxiety and depression in patients with life-threatening cancer: a randomized controlled trial. *Journal of Psychopharmacology*, 30(12): 1165–1180.

2. Swift, T. C. et al (2017). Cancer at the dinner table: experiences of psilocybin-assisted psychotherapy for the treatment of cancer-related distress. *Journal of Humanistic Psychology*, 57(5): 488–519.

3. Carhart-Harris, R. L. et al (2016). Psilocybin with psychological support for treatment-resistant depression: an open-label feasibility study. *Lancet Psychiatry*, 3(7): 619–627.

4. Carhart-Harris, R. L. et al (2017). Psilocybin with psychological support for treatment-resistant depression: six-month follow-up. *Psychopharmacology*, 235(2): 399–408.

5. Roseman, L. et al (2018). Quality of acute psychedelic experience predicts therapeutic efficacy of psilocybin for treatment-resistant depression. *Frontiers in Pharmacology*, https://doi.org/10.3389/fphar.2017.00974, published online 17 January 2018, accessed 7 March 2019.

6. Kang, M-Y. et al (2017). The relationship between shift work and mental health among electronics workers in South Korea: a cross-sectional study. *PLOS ONE*, https://doi.org/10.1371/journal.pone.0188019, published online 16 November 2017, accessed 7 March 2019.

7. Walker, M. (2017). *Why We Sleep: The New Science of Sleep and Dreams.* Penguin, UK.

8. Rechtschaffen, A. et al (1989). Sleep deprivation in the rat: X. Integration and discussion of the findings. *Sleep*, 12(1): 68–87.

9. Gregory, R. L. (1980). Perceptions as hypotheses. *Philosophical Transactions of the Royal Society London B*, 290: 181–197.

10. Clark, A. (2013). Whatever next? Predictive brains, situated agents, and the future of cognitive science. *Behavioral and Brain Sciences*, 36: 181–253.

11. Terhune, D. B. et al (2017). Hypnosis and top-down regulation of consciousness. *Neuroscience & Biobehavioral Reviews*, 81(A): 59–74.

12. Benson, H. et al (1982). Body temperature changes during the practice of g Tum-mo yoga. *Nature*, 295: 234–236.

13. Cram, A. (16 July 2007). North Pole swimmer's unique body heat trick. *The Telegraph*.

## 2: *Dream On*

1. Luby, E. D. et al (1960). Sleep deprivation: effects on behaviour, thinking, motor performance, and biological energy transfer systems. *Psychosomatic Medicine*, XXII(3): 182–192.

2. Petrovsky, N. et al (2014). Sleep deprivation disrupts prepulse inhibition and induces psychosis-like symptoms in healthy humans. *Journal of Neuroscience*, 34(27): 9134–9140.

3. Waters, F. et al (2018). Severe sleep deprivation causes hallucinations and a gradual progression toward psychosis with increasing time awake. *Frontiers in Psychiatry*, https://doi.org/10.3389/fpsyt.2018.00303, published online 10 July 2018, accessed 7 March 2019.

4. van der Helm, E. et al (2011). REM sleep depotentiates amygdala activity to previous emotional experiences. *Current Biology*, 21(23): 2029–2032.

5. Walker, M. (2017). *Why We Sleep: The New Science of Sleep and Dreams*. Penguin, UK, pp. 211–214.

6. Raskind, M. A. et al (2003). Reduction of nightmares and other PTSD symptoms in combat veterans by prazosin: a placebo-controlled study. *The American Journal of Psychiatry*, 160(2): 371–373.

7. Gujar, N. et al (2011). A role for REM sleep in recalibrating the sensitivity of the human brain to specific emotions. *Cerebral Cortex*, 21(1): 115–123.

8. Walker, M. (2017). *Why We Sleep: The New Science of Sleep and Dreams*. Penguin, UK, p. 309.

9. Hobson, J. A., Hong, C. C. and Friston, K. J. (2014). Virtual reality and consciousness inference in dreaming. *Frontiers in Psychology*, https://doi.org/10.3389/fpsyg.2014.01133, published online 9 October 2014, accessed 7 March 2019.

10. Tononi, G. and Cirelli, C. (2006). Sleep function and synaptic homeostasis. *Sleep Medicine Reviews*, 10: 49–62.

11. Spoormaker, V. I. et al (2011). Effects of rapid eye movement sleep deprivation on fear extinction recall and prediction error signaling. *Human Brain Mapping*, 33: 2362–2376.

12. Rechtschaffen, A. et al (1989). Sleep deprivation in the rat: X. Integration and discussion of the findings. *Sleep*, 12(1): 68–87.

13. Hobson, J. A. (2009). REM sleep and dreaming: towards a theory of protoconsciousness. *Nature Reviews Neuroscience*, 10: 803–813.

14. Barrett, D. and McNamara, P. (2012). *Encyclopedia of Sleep and Dreams (Volume 2)*. Greenwood, Santa Barbara, California.

15. Mota-Rolim, S. A. and Araujo, J. F. (2013). Neurobiology and clinical implications of lucid dreaming. *Medical Hypotheses*, 81(5): 751–756.

16. Hearne, K. M. T. (1978). Lucid dreams: an electro-physiological and psychological study. [PhD thesis] Department of Psychology, University of Liverpool.

17. Voss, U. (2014). Lucid dreaming: hacking the unconscious mind. [Public lecture] https://www.youtube.com/watch?v=gofc3n5biSA, accessed 7 March 2019.

18. Voss, U. et al (2009). Lucid dreaming: a state of consciousness with features of both waking and non-lucid dreaming sleep. *Sleep*, 32(9): 1191–1200.

19. Dresler, D. et al (2012). Neural correlates of dream lucidity obtained from contrasting lucid versus non-lucid REM sleep: a combined EEG/fMRI case study. *Sleep*, 35(7): 1017–1020.

20. Filevich, E. (2015). Metacognitive mechanisms underlying lucid dreaming. *The Journal of Neuroscience*, 35(3): 1082–1088.

21. Dresler, M. et al (2014). Volitional components of consciousness vary across wakefulness, dreaming and lucid dreaming. *Frontiers in Psychology*, https://doi.org//10.3389/fpsyg.2013.00987, published online 2 January 2014, accessed 7 March 2019.

22. Hobson, A. and Voss U. (2011). A mind to go out of: reflections on primary and secondary consciousness. *Consciousness and Cognition*, 20(4): 993–997.

23. Smigielski, L. et al (2019). Psilocybin-assisted mindfulness training modulates self-consciousness and brain default mode network connectivity with lasting effects. *NeuroImage*, https://doi.org/10.1016/j.neuroimage.2019.04.009, published online 6 April 2019, accessed 13 April 2019.

24. Voss, U. et al (2014). Induction of self awareness in dreams through frontal low current stimulation of gamma activity. *Nature Neuroscience*, 17: 810–812.

25. Klimke, A. et al (2016). Case report: successful treatment of therapy-resistant OCD with application of transcranial alternating current stimulation (tACS). *Brain Stimulation*, 9(3): 463–465.

26. Lutz, A. et al (2004). Long-term meditators self-induce high-amplitude gamma synchrony during mental practice. *Proceedings of the National Academy of Sciences*, 101(46): 16369–16373.

27. Arnold-Forster, M. (1921). *Studies in Dreams*. The Macmillan Company, New York.

28. Voss, U. et al (2012). Lucid dreaming: an age-dependent brain dissociation. *Journal of Sleep Research*, 21: 634–642.

29. Aspy, D. J. et al (2017). Reality testing and the mnemonic induction of lucid dreams: findings from the National Australian Lucid Dream Induction Study. *Dreaming*, 27(3): 206–231.

30. Stumbrys, T. et al (2015). Meta-awareness during day and night: the relationship between mindfulness and lucid dreaming. *Imagination, Cognition and Personality: Consciousness in Theory, Research and Clinical Practice*, 34(4): 415–433.

31. LaBerge, S. et al (2018). Pre-sleep treatment with galantamine stimulates lucid dreaming: a double-blind, placebo-controlled, crossover study. *PLOS ONE*, https://doi.org/10.1371/journal.pone.0201246, published online 8 August 2018, accessed 9 March 2019.

## 3: *Holidays from Reality*

1. BBC News (2 March 2018). UK gaming sales defy spending squeeze. http://www.bbc.co.uk/news/business-43247810, accessed 7 March 2019.

2. Blizzard Entertainment (22 January 2008). World of Warcraft reaches new milestone: 10 million subscribers. https://bit.ly/2NMAPzr, accessed 7 March 2019.

3. Jones, C. M. et al (2014). Gaming well: links between videogames and flourishing mental health. *Frontiers in Psychology*, https://doi.org/10.3389/fpsyg.2014.00260, published online 31 March 2014, accessed 7 March 2019.

4. GameTrack (ISFE/Ipsos Connect). *GameTrack Digest: Quarter 4 2016*. http://bit.ly/2GiIukj, accessed 7 March 2019.

5. Zhang, J.-T. and Brand, M. (2018). Editorial: neural mechanisms underlying internet gaming disorder. *Frontiers in Psychiatry*, https://doi.org/10.3389/fpsyt.2018.00404, published online 6 September 2018, accessed 8 March 2019.

6. van Rooij, A. J. et al (2018). A weak scientific basis for gaming disorder: Let us err on the side of caution. *Journal of Behavioral Addictions*, https://psyarxiv.com/kc7r9, accessed 8 March 2019.

7. Snodgrass, J. G. et al (2011). Magical flight and monstrous stress: technologies of absorption and mental wellness in Azeroth. *Culture, Medicine, and Psychiatry*, 35(1): 26–62.

8. Jones, C. M. et al (2014). Gaming well: links between videogames and flourishing mental health. *Frontiers in Psychology*, 5: 260.

9. MacLean, K. A. et al (2010). Intensive meditation training improves perceptual discrimination and sustained attention. *Psychological Science*, 21(6): 829–839.

10. Anguera, J. A. et al (2014). Video game training enhances cognitive control in older adults. *Nature*, 501(7465): 97–101.

11. Gackenbach, J. (2006). Video game play and lucid dreams: implications for the development of consciousness. *Dreaming*, 16(2): 96–110.

12. Gackenbach, J. I., Swanston, D. and Stark, H. (2016). Effects of video game play versus meditation/prayer in waking and dreaming experiences. *Journal of Transpersonal Psychology*, 47(2): 1–31.

13. Klasen, M. et al (2012). Neural contributions to flow experience during video game playing. *Social Cognitive and Affective Neuroscience*, 7(4): 485–495.

14. Bos, E. M. et al (2016). Out-of-body experience during awake craniotomy. *World Neurosurgery*, 92: 586.

15. Blackmore, S. J. (2017). *Seeing Myself: The New Science of Out-of-body Experiences*. Robinson, London.

16. Parnia, P. et al (2014). AWARE – AWAreness during REsuscitation – A prospective study. *Resuscitation*, 85(12): 1799–1805.

17. Ionta, S. et al (2011). Multisensory mechanisms in temporo-parietal cortex support self-location and first-person perspective. *Neuron*, 70(2): 363–374.

18. Rheingold, H. (1992). *Virtual Reality: Exploring the Brave New Technologies*. Simon & Schuster, New York.

19. Freeman, D. et al (2017). Virtual reality in the assessment, understanding, and treatment of mental health disorders. *Psychological Medicine*, 47: 2393–2400.

20. Freeman, D. et al (2018). Automated psychological therapy using immersive virtual reality for treatment of fear of heights: a single-blind, parallel group, randomised controlled trial. *The Lancet Psychiatry*, 5: 625–632.

21. Pot-Kolder, R. M. C. A. et al (2018). Virtual-reality-based cognitive behavioural therapy versus waiting list control for paranoid ideation and social avoidance in patients with psychotic disorders: a single-blind randomised controlled trial. *The Lancet Psychiatry*, 5(3): 217–226.

22. Thorens, G. et al. (2016). Capitalizing upon the attractive and addictive properties of massively multiplayer online role-playing games to promote wellbeing. *Frontiers in Psychiatry*, https://doi.org/10.3389/fpsyt.2016.00167, published online 17 October 2016, accessed 8 March 2019.

23. Tárrega, S. et al. (2015). A serious videogame as an additional therapy tool for training emotional regulation and impulse control in severe gambling disorder. *Frontiers in Psychology*, https://doi.org/10.3389/fpsyg.2015.01721, published online 12 November 2015, accessed 8 March 2019.

24. Gackenbach, J., Ellerman, E. and Hall, C. (2011). Video game play as nightmare protection: A preliminary inquiry with military gamers. *Dreaming*, 21(4), 221–245.

25. Gackenbach, J. and Hunt, H. T. (2014). 'A deeper inquiry into the association between lucid dreams and video game play'. In *Lucid Dreaming: New Perspectives on Consciousness in Sleep* edited by Ryan Hurd and Kelly Bulkeley, Praeger, Santa Barbara, California, pp. 231–254.

26. Snodgrass, J. G. et al (2011). Magical flight and monstrous stress: technologies of absorption and mental wellness in Azeroth. *Culture, Medicine, and Psychiatry*, 35(1): 26–62.

## 4: *Puppets on a String*

1. BBC (1969). The Miracle of Bali (Night). [Video] http://bbc.in/2DCopjZ, accessed 8 March 2019.

2. Dienes, Z. and Hutton, S. (2013). Understanding hypnosis metacognitively: rTMS applied to left DLPFC increases hypnotic suggestibility. *Psychiatry Research*, 49(2): 386–392.

3. Schjoedt, U. et al (2011). The power of charisma – perceived charisma inhibits the frontal executive network of believers in intercessory prayer. *Social Cognitive and Affective Neuroscience*, 6(1): 119–127.

4. Terhune, D. B. et al (2017). Hypnosis and top-down regulation of consciousness. *Neuroscience & Biobehavioral Reviews*, 81 (part A): 59–74.

5. Walsh, E. et al (2015). The functional connectivity and anatomy of thought insertion and alien control of movement. *Cortex*, 64: 380–393.

6. Röder, C. H. et al (2007). Pain response in depersonalization: a functional imaging study using hypnosis in healthy subjects. *Psychotherapy and Psychosomatics*, 76: 115–121.

7. Kosslyn, S. et al (2000). Hypnotic visual illusion alters colour processing in the brain. *The American Journal of Psychiatry*, 157(8): 1279–1284.

8. Terhune, D. B. et al (2017). Hypnosis and top-down regulation of consciousness. *Neuroscience & Biobehavioral Reviews*, 81 (part A): 59–74.

9. Schachter, S. and Singer, J. (1962). Cognitive, social, and physiological determinants of emotional state. *Psychological Review*, 69(5): 379–399.

10. Seth, A. K. (2013). Interoceptive inference, emotion, and the embodied self. *Trends in Cognitive Sciences*, 17(11): 565–573.

11. Friston, K. A. (2010). Action and behaviour: a free-energy formulation. *Biological Cybernetics*, 102(3): 227–260.

12. Rechtschaffen, A. et al (1989). Sleep deprivation in the rat: X. Integration and discussion of the findings. *Sleep*, 12(1): 68–87.

13. Moseley, G. L. et al (2008). Psychologically induced cooling of a specific body part caused by the illusory ownership of an artificial counterpart. *Proceedings of the National Academy of Sciences*, 105(35): 13169–13173.

14. David-Néel, A. (1932). *Magic and Mystery in Tibet*. Dover, New York, p. 227.

15. Benson, H. et al (1982). Body temperature changes during the practice of g Tum-mo yoga. *Nature*, 295: 234–236.

16. Bell, V. et al (2011). Dissociation in hysteria and hypnosis: evidence from cognitive neuroscience. *Journal of Neurology, Neurosurgery & Psychiatry*, 82: 332–339.

17. Seligman, R. and Kirmayer, L. (2008). Dissociative experience and cultural neuroscience: narrative, metaphor and mechanism. *Culture, Medicine and Psychiatry*, 32(1): 31–64.

18. Granqvist, P. et al (2005). Sensed presence and mystical experiences by suggestibility, not by the application of transcranial weak magnetic fields. *Neuroscience Letters*, 379: 1–6.

19. van Elk, M. and Aleman, A. (2017). Brain mechanisms in religion and spirituality: a predictive processing framework. *Neuroscience and Biobehavioral Reviews*, 73: 359–378.

20. Eason, A. D. and Parris, B. A. (2018). Clinical applications of self-hypnosis: a systematic review and meta-analysis of randomized controlled trials. *Psychology of Consciousness*, http://dx.doi.org/10.1037/cns0000173, published online 29 October 2018, accessed 9 March 2019.

21. Bo, S. et al (2018). Effects of self-conditioning techniques (self-hypnosis) in promoting weight loss in patients with severe obesity: a randomized controlled trial. *Obesity*, 26(9): 1422–1429.

## 5: *Wonder Child*

1. Hofmann, A. (2009). *LSD: My Problem Child*. MAPS, Santa Cruz, California, pp 46–52.

2. Lester Grinspoon and James B. Bakalar (1979). *Psychedelic Drugs Reconsidered*. Basic Books, New York.

3. *The Substance: Albert Hofmann's LSD*. Directed by Martin Witz. Ventura Film, 2011, https://www.youtube.com/watch?v=Nk5eYF50xII, accessed 8 March 2019.

4. Krebs, T. S. and Johansen, P-Ø. (2013). Over 30 million psychedelic users in the United States. *F1000Research*, 2: 98.

5. Gasser, P. et al (2014). Safety and efficacy of lysergic acid diethylamide-assisted psychotherapy for anxiety associated with life-threatening diseases. *The Journal of Nervous and Mental Disease*, 202(7): 513–520.

6. Ross, S. et al (2016). Rapid and sustained symptom reduction following psilocybin treatment for anxiety and depression in patients with life-threatening cancer: a randomized controlled trial. *Journal of Psychopharmacology*, 30(12): 1165–1180.

7. Griffiths, R. R. et al (2016). Psilocybin produces substantial and sustained decreases in depression and anxiety in patients with life-threatening cancer: A randomized double-blind trial. *Journal of Psychopharmacology*, 30(12): 1181–1197.

8. Carhart-Harris, R. L. et al (2016). Psilocybin with psychological support for treatment-resistant depression: an open-label feasibility study. *Lancet Psychiatry*, 7(3): 619–627.

9. Johnson, M. W. et al (2016). Long-term follow-up of psilocybin-facilitated smoking cessation. *The American Journal of Drug and Alcohol Abuse*, http://dx.doi.org/10.3109/00952990.2016.1170135, published online 21 July 2016, accessed 8 March 2019.

10. Fábregas, J. M. et al (2010). Assessment of addiction severity among ritual users of ayahuasca. *Drug and Alcohol Dependence*, 111(3): 257–261.

11. Soler, J. et al (2016). Exploring the therapeutic potential of ayahuasca: acute intake increases mindfulness-related capacities. *Psychopharmacology*, 233(5): 823–829.

12. Smigielski, L. et al (2019). Psilocybin-assisted mindfulness training modulates self-consciousness and brain default mode network connectivity with lasting effects. *NeuroImage*, https://doi.org/10.1016/j.neuroimage.2019.04.009, published online 6 April 2019, accessed 13 April 2019.

13. Carhart-Harris, R. et al (2016). Neural correlates of the LSD

experience revealed by multimodal neuroimaging. *Proceedings of the National Academy of Sciences*, 113 (17): 4853–4858.

14. Jensen, O. and Mazaheri, A. (2010). Shaping functional architecture by oscillatory alpha activity: gating by inhibition. *Frontiers in Human Neuroscience*, https://doi.org/10.3389/fnhum.2010.00186, published online 4 November 2010, accessed 8 March 2019.

15. Fair, D. A. et al (2008). The maturing architecture of the brain's default network. *Proceedings of the National Academy of Sciences*, 105(10): 4028–4032.

16. Dunbar, R. (2014). *Human Evolution*. Penguin, London, pp. 59–66.

17. Carhart-Harris, R. et al (2014) The entropic brain: a theory of conscious states informed by neuroimaging research with psychedelic drugs. *Frontiers in Neuroscience*, http://dx.doi.org/10.3389/fnhum.2014.00020, published online 3 February 2014, accessed 8 March 2019.

18. Hobson, J. A., Hong, C. C. and Friston, K. J. (2014). Virtual reality and consciousness inference in dreaming. *Frontiers in Psychology*, 5: 1133.

19. Tellegen A. and Atkinson G. (1974). Openness to absorbing and self-altering experiences ('absorption'), a trait related to hypnotic susceptibility. *Journal of Abnormal Psychology*, 83: 268–277.

20. Studerus, E. et al (2012). Prediction of psilocybin response in healthy volunteers. *PLOS ONE*, https://doi.org/10.1371/journal.pone.0030800, published online 17 February 2012, accessed 8 March 2019.

21. Ott, U. et al (2005). Evidence for a common biological basis of the absorption trait, hallucinogen effects, and positive symptoms: epistasis between 5-HT2a and COMT polymorphisms. *American Journal of Medical Genetics Part B (Neuropsychiatric Genetics)*, 137B(1): 29–32.

22. Ohayon, M. M. (2000). Prevalence of hallucinations and their pathological associations in the general population. *Psychiatry Research*, 97: 153–164.

23. Sacks, O. (2013). *Hallucinations*. Picador, London, pp. 3–44.

24. Merckelbach, H. and van de Ven, V. (2001). Another White Christmas: fantasy proneness and reports of 'hallucinatory experiences' in undergraduate students. *Journal of Behavior Therapy and Experimental Psychiatry*, 32: 137–144.

25. Corlett, P. R., Frith, C. D. and Fletcher, P. C. (2009). From drugs to deprivation: a Bayesian framework for understanding models of psychosis. *Psychopharmacology*, 206(4): 515–530.

26. Huxley, A. (2004). *The Doors of Perception*. Vintage, London, pp. 7–11.

27. Johnson, M. W. et al (2018). The abuse potential of medical psilocybin according to the 8 factors of the Controlled Substances Act. *Neuropharmacology*, 142: 143–166.

28. Hofmann, A. (2009). *LSD: My Problem Child*. MAPS, Santa Cruz, California, p. 32.

## 6: *Mother Ayahuasca*

1. Merikangas, K. R. et al (2007). Lifetime and 12-month prevalence of bipolar spectrum disorder in the National Comorbidity Survey Replication. *Archives of General Psychiatry*, 64(5): 543–552.

2. Pompili, M. (2013). Epidemiology of suicide in bipolar disorders: a systematic review of the literature. *Bipolar Disorders*, 15(5): 457–490.

3. Smith, P. (2017). What is the legality of ayahuasca in your home country? The Third Wave, https://thethirdwave.co/legality-ayahuasca/, accessed 8 March 2019.

4. Palhano-Fontes, F. et al (2018). Rapid antidepressant effects of ayahuasca in treatment-resistant depression: a randomised placebo-controlled trial. *Psychological Medicine*, https://doi.org/10.1017/S0033291718001356, published online 15 June 2018, accessed 8 March 2019.

5. dos Santos, R. G. et al (2017). Ayahuasca, dimethyltryptamine, and psychosis: a systematic review of human studies. *Therapeutic Advances in Psychopharmacology*, 7(4): 141–157.

6. Griffiths, J. (2007). *Wild: An Elemental Journey*. Hamish Hamilton, London [Kindle edition].

7. Williams, R. S. B. et al (2002). A common mechanism of action for three mood-stabilising drugs. *Nature*, 417: 292–295.

8. Saiardi, A. and Mudge, A. (2018). Lithium and fluoxetine regulate the rate of phosphoinositide synthesis in neurons: a new view of their mechanisms of action in bipolar disorder. *Translational Psychiatry*, 8(1): 175.

9. Carhart-Harris, R. L. et al (2016). Psilocybin with psychological support for treatment-resistant depression: an open-label feasibility study. *Lancet Psychiatry*, 7(3): 619–627.

10. Carhart-Harris, R. L. and Goodwin, G. M. (2017). The therapeutic potential of psychedelic drugs: past, present, and future. *Neuropsychopharmacology*, 42: 2105–2113.

11. Domínguez-Clavé, E. et al (2016). Ayahuasca: pharmacology, neuroscience and therapeutic potential. *Brain Research Bulletin*, 126(1): 89–101.

12. Nichols, D. E. (2018). Dark classics in chemical neuroscience: lysergic acid diethylamide (LSD). *ACS Chemical Neuroscience*, https://pubs.acs.org/doi/10.1021/acschemneuro.8b00043, published online 20 February 2018, accessed 8 March 2019.

13. Halberstadt, A. L. (2017). Hallucinogenic drugs: a new study answers old questions about LSD. *Current Biology*, 27(4): R156–R158.

14. Carhart-Harris, R. L. and Nutt, D. J. (2017). Serotonin and brain function: a tale of two receptors. *Journal of Psychopharmacology*, 31(9): 1091–1120.

15. Nutt, D. (2008). Equasy – an overlooked addiction with implications for the current debate on drug harms. *Journal of Psychopharmacology*, 23(1): 3–5.

16. Nutt, D. J. et al (2007). Development of a rational scale to assess the harm of drugs of potential misuse. *The Lancet*, 369(9566): 1047–1053.

17. Treiser, S. L. et al (1990). Lithium increases serotonin release and

decreases serotonin receptors in the hippocampus. *Science*, 213(4515): 1529–1531.

18. Alda, M. (2015). Lithium in the treatment of bipolar disorder: pharmacology and pharmacokinetics. *Molecular Psychiatry*, 20(6): 661–670.

## 7: *Death of the Ego*

1. Kaelen, M. et al (2015). LSD enhances the emotional response to music. *Psychopharmacology*, 232(19): 3607–3614.

2. Kaelen, M. et al (2018). The hidden therapist: evidence for a central role of music in psychedelic therapy. *Psychopharmacology*, 235(2): 505–519.

3. Kaelen, M. et al (2016). LSD modulates music-induced imagery via changes in parahippocampal connectivity. *European Neuropsychopharmacology*, 26(7): 1099–1109.

4. Carhart-Harris, R. L. et al (2015). LSD increases suggestibility in healthy volunteers. *Psychopharmacology*, 232(4): 785–794.

5. Lebedev, A. V. et al (2016). LSD-induced entropic brain activity predicts subsequent personality change. *Human Brain Mapping*, 37(9): 3203–3213.

6. Kaelen, M. (2017). The psychological and human brain effects of music in combination with psychedelic drugs. PhD thesis, Imperial College London, https://bit.ly/2t6X5Li, accessed 8 March 2019.

7. Dunbar, R. (2014). *Human Evolution*. Penguin, London, pp. 207–214.

8. Levitin, D. (2008). *This is Your Brain on Music: Understanding a Human Obsession*. Atlantic Books, London, pp. 258–260.

9. Patel, A. D. (2003). Language, music, syntax and the brain. *Nature Neuroscience*, 6(7): 674–681.

10. Fadiga, L. et al (2009). Broca's area in language, action, and music. *Annals of the New York Academy of Sciences*, 1169: 448–458.

11. Dobkin de Rios, M. (2009). *The Psychedelic Journey of Marlene*

*Dobkin de Rios: 45 Years with Shamans, Ayahuasqueros, and Ethnobotanists.* Park Street Press, Rochester, Vermont.

12. Barrett, F. S. et al (2017). Qualitative and quantitative features of music reported to support peak mystical experiences during psychedelic therapy sessions. *Frontiers in Psychology*, 8: 1238.

13. Pahnke, W. (1963). Drugs and mysticism: an analysis of the relationship between psychedelic drugs and the mystical consciousness. PhD thesis, Harvard University, Cambridge, Massachusetts.

14. St Teresa of Ávila (1946). *Interior Castle*. Tr. E. Allison Peers, Sheed & Ward, New York.

15. Julian of Norwich (1998). *Revelations of Divine Love*. Penguin Classics, London.

16. Doblin, R. (1991). Pahnke's 'Good Friday Experiment': a long-term follow-up and methodological critique. *The Journal of Transpersonal Psychology*, 23(1): 1–28.

17. Barrett, F. S. and Griffiths, R. R. (2017). Classic hallucinogens and mystical experiences: phenomenology and neural correlates. In *Behavioral Neurobiology of Psychedelic Drugs, vol. 36* (2018), edited by Adam Halberstadt et al. Springer, Berlin, pp. 393–430.

18. Devlin, H. (8 July 2017). Religious leaders get high on magic mushrooms ingredient – for science. *Guardian*. https://bit.ly/2tteH49, accessed 10 March 2019.

19. Garcia-Romeu, A. et al (2015). Psilocybin-occasioned mystical experiences in the treatment of tobacco addiction. *Current Drug Abuse Reviews*, 7(3): 157–164.

20. Ross, S. et al (2016). Rapid and sustained symptom reduction following psilocybin treatment for anxiety and depression in patients with life-threatening cancer: a randomized controlled trial. *Journal of Psychopharmacology*, 30(12): 1165–1180.

21. MacLean, K. A. et al (2011). Mystical experiences occasioned by the hallucinogen psilocybin lead to increases in the personality domain of openness. *Journal of Psychopharmacology*, 25(11): 1453–1461.

22. Lebedev, A. V. et al (2015). Finding the self by losing the self: neural correlates of ego-dissolution under psilocybin. *Human Brain Mapping*, 36(8): 3137–3153.

23. Letheby, C. and Gerrans, P. (2017). Self unbound: ego dissolution in psychedelic experience. *Neuroscience of Consciousness*, 2017(1): 1–11.

24. Millière, R. (2017). Looking for the self: phenomenology, neurophysiology and philosophical significance of drug-induced ego dissolution. *Frontiers in Human Neuroscience*, https://doi.org/10.3389/fnhum.2017.00245, published online 23 May 2017, accessed 10 March 2019.

25. Gerrans, P. and Letheby, C. (8 August 2017). Model hallucinations. *Aeon*, https://bit.ly/2vgvkRO, accessed 10 March 2019.

## 8: *The Wonderful Lightness of Being*

1. Badiner, A. and Grey, A. (2015). *Zig Zag Zen: Buddhism and Psychedelics*. Synergetic Press, Santa Fe, New Mexico.

2. Barbosa, P. C. R. et al (2018). Assessment of alcohol and tobacco use disorders among religious users of ayahuasca. *Frontiers in Psychiatry*, https://doi.org/10.3389/fpsyt.2018.00136, published online 24 April 2018, accessed 10 March 2019.

3. Halpern, J. H. et al (2008). Evidence of health and safety in American members of a religion who use a hallucinogenic sacrament. *Medical Science Monitor*, 14(8): SR15–SR22.

4. Fábregas J. M. et al (2010). Assessment of addiction severity among ritual users of ayahuasca. *Drug Alcohol Dependence*, 111(3): 257–261.

5. Thomas, G. et al (2013). Ayahuasca-assisted therapy for addiction: results from a preliminary observational study in Canada. *Current Drug Abuse Reviews*, 6(1): 30–42.

6. Bogenschutz, M. P. et al (2015). Psilocybin-assisted treatment for alcohol dependence: a proof-of-concept study. *Journal of Psychopharmacology*, 29(3): 289–299.

7. Krebs, T. S. and Johansen, P. Ø. (2012). Lysergic acid diethylamide

(LSD) for alcoholism: meta-analysis of randomized controlled trials. *Journal of Psychopharmacology*, 26(7): 994–1002.

8. Soler, J. (2016). Exploring the therapeutic potential of ayahuasca: acute intake increases mindfulness-related capacities. *Psychopharmacology*, 233(5): 823–829.

9. Soler, J. (2018). Four weekly ayahuasca sessions lead to increases in 'acceptance' capacities: a comparison study with a standard 8-week mindfulness training program. *Frontiers in Pharmacology*, https://doi.org/10.3389/fphar.2018.00224, published online 20 March 2018, accessed 10 March 2019.

10. Uthaug, M. V. et al (2018). Sub-acute and long-term effects of ayahuasca on affect and cognitive thinking style and their association with ego dissolution. *Psychopharmacology*, https://doi.org/10.1007/s00213-018-4988-3, published online 13 August 2018, accessed 10 March 2019.

11. Griffiths, R. R. et al (2018). Psilocybin-occasioned mystical-type experience in combination with meditation and other spiritual practices produces enduring positive changes in psychological functioning and in trait measures of prosocial attitudes and behaviors. *Journal of Psychopharmacology*, 32(1): 49–69.

12. Barrett, F. S. and Griffiths, R. R. (2017). Classic hallucinogens and mystical experiences: phenomenology and neural correlates. In *Behavioral Neurobiology of Psychedelic Drugs, vol. 36* (2018), edited by Adam Halberstadt et al. Springer, Berlin.

13. Kuyken, W. et al (2016). Efficacy of mindfulness-based cognitive therapy in prevention of depressive relapse: an individual patient data meta-analysis from randomized trials. *JAMA Psychiatry*, 73(6): 565–574.

14. Carhart-Harris, R. L. et al (2016). Psilocybin with psychological support for treatment-resistant depression: an open-label feasibility study. *The Lancet Psychiatry*, 3(7): 619–627.

15. Easwaran, E. (2011). *The Dhammapada* (2nd edition). The Blue Mountain Center of Meditation, Tomales, California, p. 105.

16. Seth, A. and Friston, K. (2016). Active inference and the emotional brain. *Philosophical Transactions of the Royal Society B*, 371: 20160007.

17. Pezzulo, G. et al (2015). Active inference, homeostatic regulation and adaptive behavioural control. *Progress in Neurobiology*, 134: 17–35.

18. Ondobaka, S. et al (2017). The role of interoceptive inference in theory of mind. *Brain and Cognition*, 112: 64–68.

19. Brasington, L. (2015). *Right Concentration: A Practical Guide to the Jhānas*. Shambhala, Boston and London.

20. Gunaratana, H. (7 October 2015). The taste of liberation: the jhanas. *Lion's Roar*.

21. Mumford, D. (1992). On the computational architecture of the neocortex. *Biological Cybernetics*, 66: 241–251.

22. Hagerty, M. R. et al (2013). Case study of ecstatic meditation: fMRI and EEG evidence of self-stimulating a reward system. *Neural Plasticity*, 2013: 1–12.

23. Schultz, W. (2016). Dopamine reward prediction error coding. *Dialogues in Clinical Neuroscience*, 18(1): 23–32.

24. Brasington, L. (23 May 2017). Entering the Jhanas. *Lion's Roar*.

25. Amaro, A. (2015). Meditation primer – pain. Amaravati Buddhist Monastery, www.amaravati.org/audio/meditation-primer-pain, accessed 10 March 2019.

26. Büchel, C. et al (2014). Placebo analgesia: a predictive processing framework. *Neuron*, 81(6): 1223–1239.

27. May, L. M. et al (2018). Enhancement of meditation analgesia by opioid antagonist in experienced meditators. *Psychosomatic Medicine*, 80(9): 807–813.

28. Gard, T. et al (2012). Pain attenuation through mindfulness is associated with decreased cognitive control and increased sensory processing in the brain. *Cerebral Cortex*, 22(11): 2692–2702.

29. Lindahl, J. R. et al (2017). The varieties of contemplative experience: a mixed-methods study of meditation-related challenges

in Western Buddhists. *PLOS ONE*, https://doi.org/10.1371/journal.pone.0176239, published online 24 May 2017, accessed 10 March 2019.

30. Hobson, J. A. et al (2014). Virtual reality and consciousness inference in dreaming. *Frontiers in Psychology*, https://doi.org/10.3389/fpsyg.2014.01133, published online 9 October 2014, accessed 10 March 2019.

31. Rees, A. (8 August 2004). Nobel Prize genius Crick was high on LSD when he discovered the secret of life. *Mail on Sunday*.

32. Harman, W. W. et al (1966). Psychedelic agents in creative problem-solving: a pilot study. *Psychological Reports*, 19(1): 211–227.

33. Harman, W. W. and Fadiman, J. (2016). Selective enhancement of specific capacities through psychedelic training. *Semantic Scholar*, https://bit.ly/2HbAoOI, accessed 10 March 2019.

34. Polito, V. and Stevenson, D. (2018). A systematic study of microdosing psychedelics. *PsyArXiv Preprints*, https://psyarxiv.com/cw9qs/, published online 6 December 2018, accessed 10 March 2019.

35. Jamieson, G. (2015). A unified theory of hypnosis and meditation states: the interoceptive predictive coding approach. In *Hypnosis and Meditation* (2016), edited by Amir Raz and Michael Lifshitz, OUP, UK, Chapter 16.

36. Carhart-Harris, R. et al (2016). LSD enhances suggestibility in healthy volunteers. *Psychopharmacology*, 232(4): 785–794.

37. Hölzel, B. and Ott, U. (2006). Relationship between meditation depth, absorption and meditation practice: a latent variable approach. *The Journal of Transpersonal Psychology*, 38(2): 179–199.

38. Jamieson, G. (2015). A unified theory of hypnosis and meditation states: the interoceptive predictive coding approach. In *Hypnosis and Meditation* (2016), edited by Amir Raz and Michael Lifshitz, OUP, UK, Chapter 16.

39. Killingsworth, M. A. and Gilbert, D. T. (2010). A wandering mind is an unhappy mind. *Science*, 330: 932.

## 9: *The Void Between Dreams*

1. Nichols, S. et al (2018). Death and the self. *Cognitive Science*, 42(S1): 314–332.

2. Parfit, D. (1984). *Reasons and Persons*. OUP, UK, p. 281.

3. Banaji, M. R. and Greenwald, A. G. (2016). *Blindspot: Hidden Biases of Good People*. Bantam Books, New York.

4. Huxley, A. L. (1999). *Moksha: Aldous Huxley's classic writings on psychedelics and the visionary experience*. Park Street Press, Rochester, Vermont, p. 143.

5. Hofmann, A. (2009). *LSD: My Problem Child*. MAPS, Santa Cruz, California, p. 181.

6. Gasser, P. et al (2014). Safety and efficacy of lysergic acid diethylamide-assisted psychotherapy for anxiety associated with life-threatening diseases. *The Journal of Nervous and Mental Disease*, 202: 513–520.

7. Ross, S. et al (2016). Rapid and sustained symptom reduction following psilocybin treatment for anxiety and depression in patients with life-threatening cancer: a randomized controlled trial. *Journal of Psychopharmacology*, 30(12): 1165–1180.

8. Bourdin, P. et al (2017). A virtual out-of-body experience reduces fear of death. *PLOS ONE*, https://doi.org/10.1371/journal.pone.0169343, published online 9 January 2017, accessed 11 March 2019.

9. The Western Buddhist Teachers Conference with H.H. the Dalai Lama (recorded March 1993). [Video] The Meridian Trust Tibetan Cultural Film Archive, http://meridian-trust.org/video/the-western-buddhist-teachers-conference-with-h-h-the-dalai-lama-3-of-8/4, accessed 11 March 2019.

10. Müller, M. (1884). *The Upanishads, Part II*. Clarendon Press, Oxford, Third Brâhmana, verse 22.

11. Holecek, A. (2016). *Dream Yoga*. Sounds True, Boulder, Colorado.

12. Baird, B. et al (2018). Increased lucid dream frequency in long-term meditators but not following mindfulness-based stress

training. *Psychology of Consciousness: Theory, Research, and Practice*, http://dx.doi.org/10.1037/cns0000176, accessed 11 March 2019.

13. Ferrarelli, F. et al (2013). Experienced mindfulness practitioners exhibit higher parietal-occipital EEG gamma activity during NREM sleep. *PLOS ONE*, 8(8): e73417.

14. Lutz, A. et al (2004). Long-term meditators self-induce high-amplitude gamma synchrony during mental practice. *Proceedings of the National Academy of Sciences*, 101(46): 16369–16373.

15. Borjigin, J. et al (2013). Surge of neurophysiological coherence and connectivity in the dying brain. *Proceedings of the National Academy of Sciences*, 110(35): 1432–1437.

16. Chawla, L. S. et al (2009). Surges of electroencephalogram activity at the time of death: a case series. *Journal of Palliative Medicine*, 12(12): 1095–1100.

17. Strassman, R. (2001). *DMT: The Spirit Molecule*. Park Street Press, Rochester, Vermont.

18. Timmermann, C. et al (2018). DMT models the near-death experience. *Frontiers in Psychology*, https://doi.org/10.3389/fpsyg.2018.01424, published online 15 August 2018, accessed 12 March 2019.

19. Nichols, D. E. (2017). *N,N*-dimethyltryptamine and the pineal gland: separating fact from myth. *Journal of Psychopharmacology*, 32(1): 30–36.

20. Bennett, N. (2017). David Nichols – DMT and the pineal gland: facts vs fantasy. [Video] Breaking Convention (recorded 2 July 2017, uploaded 9 July 2017), https://www.youtube.com/watch?v=YeeqHUiC8Io, accessed 12 March 2019.

21. Groth-Marnat, G. and Summers, R. (1998). Altered attitudes, beliefs, and behaviors following near-death experiences. *Journal of Humanistic Psychology*, 38(3): 110–125.

22. Carhart-Harris, R. L. and Nutt, D. J. (2017). Serotonin and brain function: a tale of two receptors. *Journal of Psychopharmacology*, 31(9): 1091–1120.

23. Lemercier, C. E. and Terhune, D. B. (2018). Psychedelics and hypnosis: commonalities and therapeutic implications. *Journal of Psychopharmacology*, 32(7): 732–740.

24. Badiner, A. (2015). *Zig Zag Zen: Buddhism and Psychedelics.* Synergetic Press, Santa Fe, New Mexico, pp. 49–58.

25. Pistono, M. (19 August 2018). The new wave of psychedelics in Buddhist practice. *Lion's Roar.*

26. Sucitto, A. (2018). Dhamma and psychedelics. [Blog] Reflections: Ajahn Sucitto (10 May 2018), http://sucitto.blogspot.com/2018/05/dhamma-and-psychedelics.html, accessed 9 March 2019.

## *Epilogue*

1. Friston, K. (2012). A free energy principle for biological systems. *Entropy*, 14(11): 2100–2121.

2. Raviv, S. (13 November 2018). The genius neuroscientist who might hold the key to true AI. *Wired.*

3. Belouin, S. J. and Henningfield, J. E. (2018). Psychedelics: where we are now, why we got here, what we must do. *Neuropharmacology*, 142: 7–19.

4. ibid.

5. Barrett, F. S. and Griffiths, R. R. (2017). Classic hallucinogens and mystical experiences: phenomenology and neural correlates. In *Behavioral Neurobiology of Psychedelic Drugs, vol. 36* (2018), edited by Adam Halberstadt et al. Springer, Berlin, pp. 393–430.

6. Garcia-Romeu, A., Griffiths, R. R. and Johnson, M. W. (2015). Psilocybin-occasioned mystical experiences in the treatment of tobacco addiction. *Current Drug Abuse Review*, 7(3): 157–164.

7. Steger, M. (2013). What makes life meaningful? [Video] TEDx Talks (recorded 8 March 2013, uploaded 14 March 2013), https://www.youtube.com/watch?v=RLFVoEF2RIo, accessed 9 March 2019.

8. James, W. (2007). *The Varieties of Religious Experience: A Study in Human Nature.* Cosimo Classics, New York.

9. Jungaberle, H. et al (2018). Positive psychology in the investigation of psychedelics: a critical review. *Neuropharmacology*, 142: 179–199.

10. Haijen, E. C. H. M. et al. Predicting responses to psychedelics: a prospective study. *Frontiers in Pharmacology*, https://doi.org/10.3389/fphar.2018.00897, published online 2 November 2018, accessed 9 March 2019.

11. Haden, M. et al (2016). A public-health-based vision for the management and regulation of psychedelics. *Journal of Psychoactive Drugs*, 48(4): 243–252.

# *Acknowledgements*

I am indebted to the UK's Psychedelic Society for making the life-enhancing potential of psychedelics a reality for me and so many others, and in particular to Stefana Bosse, Biz Bliss, Amit Elan and Brooks for facilitating our journeys of wonder and self-discovery. My thanks also to Percy Garcia Lozano of Dios Ayahuasca Sanaciones in Iquitos, Peru, for his powerful medicine, ceremony and songs.

For their pioneering research and advocacy of science-based, rational drug laws, the world is indebted to Amanda Feilding and the Beckley Foundation; David Nutt, Robin Carhart-Harris and the Imperial College Psychedelic Research Group in London; Rick Doblin and MAPS (the Multidisciplinary Association for Psychedelic Studies); Gerald Thomas at the University of Victoria, British Columbia; Jordi Riba's team at the Sant Pau Biomedical Research Institute in Barcelona; Roland Griffiths, William Richards and their colleagues at Johns Hopkins University School of Medicine; David Nichols at the University of North Carolina, Chapel Hill; Benjamin Mudge at Flinders University, Adelaide; and all the other scientists and therapists who have overcome professional, social and institutional prejudice to advance this important field.

I will be eternally grateful to Karl Friston, Allan Hobson, Anil Seth, Jayne Gackenbach and Graham Jamieson for patiently explaining to me their extraordinary insights into the neuroscience of consciousness and its infinite variations.

For setting my feet on the path, I thank Bhante Henepola Gunaratana for his wonderful book *Mindfulness in Plain English*, and Ajahn Amaro, Ajahn Vimalo and Ajahn Anando at Amaravati Buddhist Monastery for all their profound teachings over the years.

Special thanks are also due to my agent Peter Tallack at the Science Factory for helping me prepare the proposal for this book, to Mike Harpley at Atlantic Books for giving me a shot at it, to editor James Pulford for winkling out the real stories behind it, and to Julia Kellaway for her meticulous copy-editing.

Last but not least, heartfelt gratitude to James Drever for his comments on the manuscript, and to my family, friends and partner Art for watching over me as I researched and wrote this book.

# *Index*